MATERIALS SCIENCE AND ENGINEERING FOR THE 1990s

Maintaining Competitiveness in the Age of Materials

Committee on Materials Science and Engineering

Solid State Sciences Committee
Board on Physics and Astronomy
Commission on Physical Sciences, Mathematics, and Resources
and
National Materials Advisory Board
Commission on Engineering and Technical Systems

National Research Council

NATIONAL ACADEMY PRESS
Washington, D.C.   1989

National Academy Press ● 2101 Constitution Avenue, NW ● Washington, DC 20418

NOTICE: The project that is the subject of this report was approved by the Governing Board of the National Research Council, whose members are chosen from the councils of the National Academy of Sciences, the National Academy of Engineering, and the Institute of Medicine. The members of the committee responsible for the report were chosen for their special competences and with regard for appropriate balance.

This report has been reviewed by a group other than the authors according to procedures approved by a Report Review Committee consisting of members of the National Academy of Sciences, the National Academy of Engineering, and the Institute of Medicine.

This report was supported by the National Aeronautics and Space Administration, the Department of Energy, the Army Research Office, the Defense Advanced Research Projects Agency, the Air Force Office of Scientific Research, and the National Science Foundation under Grant No. DMR-8521440. Additional support came from the National Research Council (NRC) Fund, a pool of private, discretionary, nonfederal funds that is used to support a program of Academy-initiated studies of national issues in which science and technology figure significantly. The NRC Fund consists of contributions from a consortium of private foundations including the Carnegie Corporation of New York, the Charles E. Culpeper Foundation, the William and Flora Hewlett Foundation, the John D. and Catherine T. MacArthur Foundation, the Andrew W. Mellon Foundation, the Rockefeller Foundation, and the Alfred P. Sloan Foundation; and from the Academy Industry Program, which seeks annual contributions from companies that are concerned with the health of U.S. science and technology and with public policy issues with technological content.

Library of Congress Cataloging-in-Publication Data

National Research Council (U.S.). Committee on Materials Science and Engineering.
   Materials science and engineering for the 1990s : maintaining competitiveness in the age of materials / Committee on Materials Science and Engineering [and] Solid State Sciences Committee, Board on Physics and Astronomy, Commission on Physical Sciences, Mathematics, and Resources, and National Materials Advisory Board, Commission on Engineering and Technical Systems, National Research Council.
     p.   cm.
   ISBN 0-309-03928-2.
   1. Materials science. 2. Engineering. I. National Research Council (U.S.). Solid State Sciences Committee. II. National Research Council (U.S.). National Materials Advisory Board. III. Title.
TA403.N332   1989
620.1'1—dc20   89-12630
                                      CIP

Cover: Computer-generated image of a mathematical model of Scherk's first minimal surface. (Reprinted, by permission, from Edwin L. Thomas, David M. Anderson, Chris S. Henkee, and David Hoffman, 1988, Periodic Area-Minimizing Surfaces in Block Copolymers, Nature 334:598–601. Copyright © 1988 by Macmillan Magazines Ltd.) Computer graphics by James T. Hoffman, Geometry, Analysis, and Numerics Group (GANG), Department of Mathematics, University of Massachusetts.

First Printing, September 1989
Second Printing, May 1990

Printed in the United States of America

This report, <u>Materials Science and Engineering for the 1990s: Maintaining Competitiveness in the Age of Materials</u>, encompasses a broad enterprise. The field's intellectual content ranges from the quantized Hall effect to dramatic advances in the performance of high-strength structural materials. The vitality and pace of the field are everywhere evident. The Nobel Committee recognized fundamental advances in materials research for three consecutive years -- 1985, 1986, and 1987. Success in translating advances in materials science and engineering into new and improved materials is unparalleled. There have been gratifyingly broad applications of new materials in areas with immediate impact on human welfare such as biomaterials, suitable for artificial organs, biochemical sensors, vascular grafts, and ophthalmological devices.

Despite the diversity of the field, the report points to unifying trends that emphasize the need for scientists and engineers in universities, government laboratories, and industry to work together closely. In particular, the authoring group, the Committee on Materials Science and Engineering, urged greater efforts by the federal government to coalesce these sectors, and endorsed Congressional efforts to strengthen the coordination of federal agencies that support materials science and engineering.

The committee focused its recommendations on synthesis and processing of materials. This is the area that has produced dramatic improvements in superconducting materials, growth in the number of components in integrated circuits, and increases in the strength of structural materials. On the basis of a survey of several key industries, the committee recommended a national initiative in synthesis and processing built on cooperation among universities, industry, and government.

We believe that the field of materials science and engineering offers a special opportunity to act on the growing realization of the need for improved coordination and cooperation in the nation's effort in science and technology. We commend the report to your attention.

Frank Press
Chairman
National Research Council

Robert M. White
Vice Chairman
National Research Council

THE NATIONAL RESEARCH COUNCIL IS THE PRINCIPAL OPERATING AGENCY OF THE NATIONAL ACADEMY OF SCIENCES AND THE NATIONAL ACADEMY OF ENGINEERING

TO SERVE GOVERNMENT AND OTHER ORGANIZATIONS.

The National Academy of Sciences is a private, nonprofit, self-perpetuating society of distinguished scholars engaged in scientific and engineering research, dedicated to the furtherance of science and technology and to their use for the general welfare. Upon the authority of the charter granted to it by Congress in 1863, the Academy has a mandate that requires it to advise the federal government on scientific and technical matters. Dr. Frank Press is president of the National Academy of Sciences.

The National Academy of Engineering was established in 1964, under the charter of the National Academy of Sciences, as a parallel organization of outstanding engineers. It is autonomous in its administration and in the selection of its members, sharing with the National Academy of Sciences the responsibility for advising the federal government. The National Academy of Engineering also sponsors engineering programs aimed at meeting national needs, encourages education and research, and recognizes the superior achievements of engineers. Dr. Robert M. White is the president of the National Academy of Engineering.

The Institute of Medicine was established in 1970 by the National Academy of Sciences to secure the services of eminent members of appropriate professions in the examination of policy matters pertaining to the health of the public. The Institute acts under the responsibility given to the National Academy of Sciences by its congressional charter to be an adviser to the federal government and, upon its own initiative, to identify issues of medical care, research, and education. Dr. Samuel O. Thier is president of the Institute of Medicine.

The National Research Council was organized by the National Academy of Sciences in 1916 to associate the broad community of science and technology with the Academy's purposes of furthering knowledge and of advising the federal government. Functioning in accordance with general policies determined by the Academy, the Council has become the principal operating agency of both the National Academy of Sciences and the National Academy of Engineering in providing services to the government, the public, and the scientific and engineering communities. The Council is administered jointly by both Academies and the Institute of Medicine. Dr. Frank Press and Dr. Robert M. White are chairman and vice chairman, respectively, of the National Research Council.

STEERING COMMITTEE FOR MATERIALS SCIENCE AND ENGINEERING

ALBERT NARATH, AT&T Bell Laboratories, *Co-chairman*
ARDEN L. BEMENT, JR., TRW, Inc., *Co-chairman*
JOHN H. BIRELY, Los Alamos National Laboratory
MORRIS COHEN, Massachusetts Institute of Technology
WALTER KOHN, University of California
WILLIAM P. SLICHTER, AT&T Bell Laboratories (*retired*)

Ex-Officio Members
PRAVEEN CHAUDHARI, IBM T. J. Watson Research Center,
   Co-chairman, Committee on Materials Science and Engineering
MERTON C. FLEMINGS, Massachusetts Institute of Technology,
   Co-chairman, Committee on Materials Science and Engineering
HERBERT H. JOHNSON, Cornell University, *Chairman*, Solid State
   Sciences Committee
BERNARD H. KEAR, Rutgers University, *Chairman*, National Materials
   Advisory Board
NORMAN F. RAMSEY, Harvard University, *Chairman*, Board on Physics
   and Astronomy

DONALD C. SHAPERO, Staff Director, Board on Physics and Astronomy
KLAUS M. ZWILSKY, Staff Director, National Materials Advisory Board

COMMITTEE ON MATERIALS SCIENCE AND ENGINEERING

PRAVEEN CHAUDHARI, IBM T. J. Watson Research Center,
   Co-chairman
MERTON C. FLEMINGS, Massachusetts Institute of Technology,
   Co-chairman
MELVIN BERNSTEIN, Illinois Institute of Technology
MARTIN BLUME, Brookhaven National Laboratory
ALAN CHYNOWETH, Morris Research & Engineering Center, Bell
   Communications Research, Inc.
W. DALE COMPTON, Purdue University
ROBERT S. HANSEN, Iowa State University
JOHN HULM, Westinghouse Electric Research and Development Center
R. GLEN KEPLER, Sandia National Laboratories
JAMES S. LANGER, University of California

v

TERRY L. LOUCKS, Rothschild Ventures
GEORGE PARSHALL, E. I. du Pont de Nemours & Co., Inc.
RUSTUM ROY, Pennsylvania State University
LYLE H. SCHWARTZ, National Institute of Standards and Technology
JAMES O. STIEGLER, Oak Ridge National Laboratory
GEORGE WHITESIDES, Harvard University
JAMES C. WILLIAMS, General Electric Company

DONALD C. SHAPERO, Staff Director, Board on Physics and Astronomy
KLAUS M. ZWILSKY, Staff Director, National Materials Advisory Board
JACK MOTEFF, NRC Fellow (1985–1987)
ARLENE MACLIN, Program Officer (1985–1987)
PATRICK RAPP, Program Officer (1988)
STEVE OLSON, Consultant (1987–1988)

Government Liaison Representatives

TED BERLINCOURT, Director, Research and Laboratory Management,
  Office of the Under Secretary of Defense for Research and Advanced
  Technology, Department of Defense
ADRIAAN de GRAAF, Division of Materials Research, National Science
  Foundation
B. CHALMERS FRAZER, Solid State Physics and Materials Chemistry,
  U.S. Department of Energy
RICHARD E. HALPERN, National Aeronautics and Space Administration
LOUIS C. IANNIELLO, Deputy Associate Director for Basic Energy
  Sciences, U.S. Department of Energy
PAUL MAXWELL, Committee on Science and Technology, U.S. House of
  Representatives
GEORGE MAYER, Director, Materials Science Division, U.S. Army
  Research Office
RICHARD REYNOLDS, Director, Defense Science Office, Defense
  Advanced Research Projects Agency
ALAN ROSENSTEIN, Air Force Office of Scientific Research
AL SCHINDLER, Director, Division of Materials Research, National
  Science Foundation
IRAN THOMAS, Director, Division of Materials Sciences, U.S.
  Department of Energy
DONALD R. ULRICH, Senior Program Manager, Chemical Sciences, Air
  Force Office of Scientific Research
ROBERT WEIGLE, U.S. Army Research Office
BEN WILCOX, Assistant Director, Materials Science Division, Defense
  Advanced Research Projects Agency

THEODORE GEBALLE, Stanford University
GORDON H. GEIGER, North Star Steel Company
FRANK E. JAMERSON, General Motors Research Laboratories
HARRY A. LIPSITT, Wright State University
JAMES L. McCALL, Battelle Columbus Division
THOMAS C. McGILL, JR., California Institute of Technology
JOHN P. RIGGS, Hoechst Celanese Corporation
GERD M. ROSENBLATT, Lawrence Berkeley Laboratory
PALLE SMIDT, Microelectronics Corporation
ROBERT STREET, Xerox Palo Alto Research Center
HILLIARD WILLIAMS, Monsanto Company

ARPAD A. BERGH, Morris Research & Engineering Center, Bell
   Communications Research, Inc., Consultant

DONALD C. SHAPERO, Staff Director, Board on Physics and Astronomy
KLAUS M. ZWILSKY, Staff Director, National Materials Advisory Board

PANEL ON INTERNATIONAL COOPERATION AND COMPETITION IN MATERIALS SCIENCE AND ENGINEERING

LYLE H. SCHWARTZ, National Institute of Standards and Technology,
   Chairman
W. DALE COMPTON, Purdue University, *Vice-Chairman*
RUSTUM ROY, Pennsylvania State University, *Vice-Chairman*
JORDAN BARUCH, Jordon Baruch Associates
C. PETER FLYNN, University of Illinois
RICHARD J. FRUEHAN, Carnegie-Mellon University
HERBERT I. FUSFELD, Rensselaer Polytechnic Institute
SERGE GRATCH, GMI Engineering and Management Institute
RUDOLPH PARISER, E. I. du Pont de Nemours & Co., Inc.
R. BYRON PIPES, University of Delaware
MAXINE SAVITZ, The Garrett Corporation
GABOR A. SOMORJAI, University of California
GREGORY STILLMAN, University of Illinois
JAMES J. TIETJEN, RCA Laboratories
ROBERT WHITE, Control Data Corporation

SAMUEL SCHNEIDER, National Institute of Standards and Technology,
   Consultant

DONALD C. SHAPERO, Staff Director, Board on Physics and Astronomy
KLAUS M. ZWILSKY, Staff Director, National Materials Advisory Board

viii

PANEL ON RESEARCH RESOURCES IN MATERIALS SCIENCE AND ENGINEERING

TERRY L. LOUCKS, Rothschild Ventures, *Chairman*
MARTIN BLUME, Brookhaven National Laboratory, *Vice-Chairman*
GEORGE WHITESIDES, Harvard University, *Vice-Chairman*
BILL R. APPLETON, Oak Ridge National Laboratory
ROBERT S. BAUER, Xerox Palo Alto Research Center
H. KENT BOWEN, Massachusetts Institute of Technology
PETER M. EISENBERGER, Exxon Research & Engineering Co.
NICHOLAS F. FIORE, Cabot Corporation
JOHN J. GILMAN, Lawrence Berkeley Laboratory
KARL HESS, University of Illinois
ISRAEL S. JACOBS, General Electric Research and Development Center
J. DAVID LITSTER, Massachusetts Institute of Technology
NOEL MacDONALD, Cornell University
DENNIS McWHAN, AT&T Bell Laboratories
EMIL PFENDER, University of Minnesota
BHAKTA B. RATH, Naval Research Laboratories
JOHN S. RYDZ, Emhart Corporation
ISAAC F. SILVERA, Harvard University
RICHARD S. STEIN, University of Massachusetts
JULIA WEERTMAN, Northwestern University

DONALD C. SHAPERO, Staff Director, Board on Physics and Astronomy
KLAUS M. ZWILSKY, Staff Director, National Materials Advisory Board

PANEL ON EDUCATION IN MATERIALS SCIENCE AND ENGINEERING

MELVIN BERNSTEIN, Illinois Institute of Technology, *Chairman*
ROBERT S. HANSEN, Iowa State University, *Vice-Chairman*
JOHN HULM, Westinghouse Electric Research and Development Center, *Vice-Chairman*
DIRAN APELIAN, Drexel University
ALI S. ARGON, Massachusetts Institute of Technology
MALCOLM R. BEASLEY, Stanford University
GILBERT Y. CHIN, AT&T Bell Laboratories
ROBERT CLAGETT, University of Rhode Island
ANTHONY G. EVANS, University of California
LEROY EYRING, Arizona State University
HELLMUT FRITZSCHE, University of Chicago
BRUCE N. HARMON, Iowa State University

Preface

In October 1984 Don Fuqua, then chairman of the House Committee on Science and Technology, wrote to the presidents of the National Academy of Sciences and the National Academy of Engineering urging the National Research Council to form a committee "to conduct a comprehensive materials research and technology assessment for the next decade." This direct expression of support from a U.S. congressman, which was further reinforced by the federal agencies with materials-related missions, marked the inception of the survey of materials science and engineering presented here. But the roots of this survey extend much further to include the initial recognition of materials science and engineering as a distinct area of endeavor. There have been earlier comprehensive studies of materials science and engineering, most notably that of the National Research Council's Committee on the Survey of Materials Science and Technology (COSMAT). The publication of COSMAT's 1975 report *Materials and Man's Needs* moved understanding and recognition of the field forward. At that time, national goals were focused on natural resources, energy, and the environment, as well as on defense. *Materials and Man's Needs* dealt with materials issues related to strategic materials, reduction of energy costs in production, biodegradability, recovery and recycling of scrap, and the materials cycle, all in the context of an awakened public awareness of the finiteness of the earth's resources. It also discussed the structure-property-performance relationships that have been so important to development of the field over the last decades. The present report, building on the foundation of that earlier report, stresses the importance of synthesis and processing.

At the inception of this survey, it was clear that materials science and engineering had changed dramatically since the completion of the COSMAT report. A wealth of new discoveries and technological advances had drawn many new people to the field and had radically altered the field's concerns and methods. At the same time, a number of industries closely associated with materials science and engineering had undergone similarly dramatic changes—and not always for the better. America's mining and metals beneficiation industries, its commodity metals industry, its machine tool industry, its computer industry, and its electronics industry, which had been, and still are, major users of the results of materials science and engineering, were all losing major portions of their market shares to overseas competitors and shutting down research operations.

Prompted by Fuqua's letter, the Solid State Sciences Committee, in collaboration with the National Materials Advisory Board, devoted its spring 1985 forum to the question of whether a new survey of materials science and engineering should be conducted and, if so, how it should be structured. At the forum a remarkable degree of unanimity emerged regarding the potential value of such a study, and forum participants outlined a general statement of task for the project. Shortly thereafter, the National Research Council initiated a joint project under the Solid State Sciences Committee and the National Materials Advisory Board to conduct a survey along the lines suggested, and funding was obtained from the National Science Foundation, the Department of Energy, and the Defense Advanced Research Projects Agency, as well as from the National Aeronautics and Space Administration, the Air Force Office of Scientific Research, the Army Research Office, and the National Research Council.

The National Research Council's principal goal for the study was to present "a unified view of recent progress and new directions in materials science and engineering." Among the specific issues identified in the charge were

- areas of research and development particularly ripe for important advances;
- relationships among the various elements of materials research and development;
- the roles of the federal and private sectors, particularly as they relate to a balanced national materials effort;
- the effectiveness of the materials infrastructure in developing and commercializing new materials technologies;
- the effectiveness of materials research and education at universities; and
- international cooperation and competition in materials science and engineering.

The Committee on Materials Science and Engineering was constituted by the National Research Council with a special focus on the unity of materials science and engineering. The committee was carefully balanced with respect to several different factors including the range of disciplines that conduct materials science and engineering, the variety of institutions in which these activities take place, and the scope of the field from science to engineering. By the summer of 1986, a committee of 17 eminently qualified individuals representing government, industry, and academia had been formed. In addition, a steering committee was established to provide oversight and guidance throughout the committee's deliberations.

One of the first and most challenging tasks facing the committee was to find a way of breaking down a subject as large and complex as materials science and engineering into manageable parts. The committee formed five panels, each of which examined an important area of the field that cut across all materials classes and ranged from science to engineering to industrial practice. The Panel on Research Opportunities and Needs in Materials Science and Engineering identified research areas of national importance in materials science and engineering and evaluated opportunities and needs in the field. The Panel on Exploitation of Materials Science and Technology for the National Welfare examined the links between scientific advances and economically competitive products and processes and other ways in which materials science and engineering affects the national well-being. The Panel on International Cooperation and Competition in Materials Science and Engineering outlined the global dimensions of the field, particularly as it affects industrial competitiveness in the United States. The Panel on Research Resources in Materials Science and Engineering assessed the resources available now and in the future for materials science and engineering in terms of facilities, instrumentation, and funding at universities, national laboratories, and industrial laboratories. The Panel on Education in Materials Science and Engineering considered personnel issues and the means by which future generations of materials scientists and engineers are to be educated.

The leadership of each panel consisted of one chairman and two vice chairmen drawn from the committee (the two committee co-chairmen were the only committee members not serving on a panel). Panel leaders included one person from industry, one from a government laboratory, and one from academia. In turn, the National Research Council appointed a balanced panel, and the panels conducted meetings and surveys, commissioned papers, and gathered data. In this way, a broad cross section of the materials community was involved in the preparation of this report (there were 109 formally constituted committee and panel members and nearly 400 other individuals who contributed to the study). The co-chairmen of the committee and the committee members also appeared before a number of professional societies to present status reports on the committee's deliberations and to encourage participation and feedback.

Each panel produced a major report on its assigned issue, and these panel reports form the basis for this report. However, this report is not organized strictly along the lines of the issues assigned to the panels. As intended, there was considerable overlap among the panels, an overlap that contributed to the richness of the committee's conclusions. This report builds on that overlap to provide a committee consensus of all of the panels' conclusions. Although the findings of particular panels may contribute more heavily to some of the chapters in this report than to others, in effect, each of the panels contributed to each of the chapters of this report. The introduction to the report briefly reviews the contents of the chapters.

We would like to mention one issue that arose from the work of the Panel on Research Opportunities and Needs in Materials Science and Engineering and the work of the Panel on International Cooperation and Competition in Materials Science and Engineering that is treated in Chapter 2. The work of these panels uncovered significant issues with regard to competitiveness. One of the committee's conclusions is that better integration of materials science and engineering with the rest of business operations is needed to improve the positions of U.S. firms in domestic and international competition; the objective is to strengthen long-range R&D in industry. Other issues of competitiveness emerged that are alluded to above and that have more to do with the entire structure and climate of industry in the United States. These issues are profound and deserve more attention than a study whose scope is limited to materials science and engineering can give them.

As would be expected for a diverse field such as materials science and engineering, the findings range over many topics, and the recommendations are broad in character. In the spirit that has characterized the whole endeavor of the Committee on Materials Science and Engineering, this report is offered with the hope that its important recommendations will be adopted and implemented in ways that will benefit the United States.

PRAVEEN CHAUDHARI
MERTON FLEMINGS
Co-chairmen
Committee on Materials Science and Engineering

Acknowledgments

Successful completion of this report involved the contributions of many.

The work of the committee was supported by National Research Council staff members Arlene Maclin, Jack Moteff, and Pat Rapp as well as Board on Physics and Astronomy and National Materials Advisory Board directors Don Shapero and Klaus Zwilsky. The editing team included Roseanne Price, Susan Maurizi, and Susan Wyatt. Writer Steve Olson's work in synthesizing the reports of the five panels was indispensable. Oversight by a steering committee co-chaired by Arden Bement and Al Narath helped at several crucial junctures along the way to completing the project. A critical review process overseen by the Report Review Committee contributed to the refining of this report. The National Academy Press staff designed the book and brought it through production.

The Committee on Materials Science and Engineering would also like to thank the following members of the materials community for their assistance in providing information for this report: Aerospace Industry Subpanel members Peter Cannon (Chairman), Donald P. Ames, Andrew Baker, Arden Bement, Wayne Burwell, Richard Delasi, Russell Duttweiller, Richard Hartke, Stephen Lukasik, Edith Martin, Robert Sprague, Earl Thompson, James Whitesides, and Carl Zweben; Automotive Industry Subpanel members C. Magee (Chairman), P. Beardmore, H. Cook, J. Hunter, M. Liedtke, A. McLean, G. Robinson, and R. Sjoberg; Biomaterials Industry Subpanel members S. Barenberg (Chairman), J. Andrade, P. Bosen, R. Crowninshield, P. Galetti, W. Grantz, A. Haubold, M. Helmus, R. Kronenthal, J. Lemmons, L. Lynch, E. Mueller, M. Ostler, M. Refojo, S. Shalaby, J. Shaw, and J. Williams; Chemistry Subpanel members George Hammond (Chairman), James

Clovis, Ted Evans, Edith Flanigen, Lawrence Hare, Harris Hartzler, Robert Jannson, Donald McLemore, and Lloyd Robeson; Electronics Industry Subpanel members Bob Stratton (Chairman), Al Cho, Dick Delagi, William Gallagher, Kent Hansen, Webb Howard, Howard Huff, Milo Johnson, Bill Mitchell, Elsa Reichmanis, Bob Rosenberg, Ralph Ruth, and Pei Wang; Energy Industry Subpanel members R. Jaffee (Chairman), E. DeMeo, B. Kear, W. Liang, R. Richman, J. Roberts, and D. Shannon; Metals Industry Subpanel members Ian Hughes (Chairman), Philip Abramowitz, Yaz Bilimoria, Larry Hicks, Noel Jarrett, John Mihelich, Neil Paton, and Joseph Winter; and Telecommunications Industry Subpanel members Robert Laudise (Chairman), Glenn Cullen, Barry Dunbridge, Kenneth Jackson, Charles Jonscher, Robert Maurer, Gregory Stillman, and Jack Wernick.

In addition, the committee would like to thank those who participated in the workshop held by the Panel on Materials Research Opportunities and Needs in Materials Science and Engineering: Harry Allcock, Sumner Barenberg, Malcolm Beasley, H. Kent Bowen, Morris Cohen, Lance Davis, Frank Di Salvo, Anthony Evans, Paul Fleury, John Hirth, John Joannopoulos, Frank Karasz, Bernard Kear, David Litster, Alex Maradudin, Robert Mehrabian, Raumond Orbach, Richard Osgood, John Quinn, John Silcox, Robert White, and James Williams.

The committee is also grateful to the participants of the two workshops held by the Panel on Exploitation of Materials Science and Technology for the National Welfare: Workshop on Technological Innovations and Technology Transfer participants Alan Chynoweth (Chairman), Michael Chartock, Joel Clark, J. William Doane, Ted Geballe, Harry Gibson, Lyman Johnson, Harry Lipsitt, Stewart Miller, Phillip Parrish, John Riggs, Palle Smidt, Robert Sundahl, and Port Wheeler; and Workshop on Institutional Aspects of Technology Transfer participants Alan Chynoweth (Chairman), Gordon Geiger, Sigfried Hecker, Herb Johnson, Ronald Kerber, Bob McKee, Richard Pitler, Vince Russo, Larry Sumney, Douglas Walgren, and Karl Willenbrock.

Thanks are also extended by the committee to those who participated in case studies, including C. Flynn, G. Somorjai, W. Dennis, H. Paxton, and L. Kuhn. We acknowledge with gratitude the help of A. Malozemoff and R. Rosenberg with writing sections of this report.

PRAVEEN CHAUDHARI
MERTON FLEMINGS
Co-chairmen
Committee on Materials Science and Engineering

Contents

APPENDIXES: ISSUES IN MATERIALS RESEARCH

MATERIALS SCIENCE AND ENGINEERING FOR THE 1990s

Summary, Conclusions, and Recommendations

SUMMARY

This study of materials science and engineering has produced a picture of remarkable contrasts. On the one hand, the study has revealed a field of great vitality—rapidly emerging scientific discoveries, stunning new capabilities for understanding and prediction, and applications that are essential for the health of every U.S. industry. On the other hand, several troubling developments have come to light. Despite growing opportunities in the field, a shortage of educated personnel is foreseen. Limitations on resources are constraining progress. And our national effort needs greater focus and coordination in order to meet the challenge of international competition.

This picture evolved as a result of the work of five panels that addressed research opportunities and needs, exploitation of materials science and engineering for the national welfare, international cooperation and competition, research resources, and education. Each of the panels submitted detailed reports to the Committee on Materials Science and Engineering. The charge to the Committee on Materials Science and Engineering was "to present a unified view of recent progress and new directions in materials science and engineering and to assess future opportunities and needs." The committee conducted this study with a view to developing the consensus implied by the phrase "unified view." The objective of cultivating this consensus in the very diverse materials science and engineering community was taken no less seriously than that of carrying out the scientific and engineering assessment contained in this report. The main conclusions are described in the seven chapters of this volume.

Chapter 1 briefly discusses the significance and development of materials science and engineering as an interdisciplinary endeavor that profoundly affects our quality of life in many ways. The potential economic and strategic impact of materials science and engineering is examined in Chapter 2 through a study of the materials needs of eight industries that collectively have sales of $1.4 trillion. Scientific and technological frontiers are explored in Chapters 3 and 4, both from the point of view of materials classes and from the point of view of the four elements of the field: synthesis and processing, structure and composition, properties, and performance. It is here that several aspects of the field become apparent: rapid progress at the forefronts, an emerging sense of unity, and a critical weakness in the area of synthesis and processing of materials. Issues related to synthesis, processing, performance, instrumentation, and analysis and modeling—areas considered essential to the progress of research in materials science and engineering—are discussed in greater detail in Appendixes A to E, respectively. In Chapter 5, which describes manpower and education in materials science and engineering, a picture of the richness of the field appears—the opportunities in the field draw physicists, chemists, biologists, and materials engineers together to solve materials problems. But the committee identified a critical need for new curricula and for increased production of educated manpower from university departments involved with materials science and engineering. Again, in assessing the resource needs discussed in Chapter 6, the committee found signs of trouble. Federal programs are shrinking rather than growing, and there is a critical need for facilities in the area of synthesis and processing. Finally, the role of materials in U.S. manufacturing success and ability to compete in global markets is treated in Chapter 7. All the major industrialized nations surveyed are revealed to have a strong commitment to industrial growth, stimulated by coordinated R&D in materials; the governments of all of these countries actively foster cooperative mechanisms to enhance competitiveness.

The central message of this report is a challenge both to the community of materials scientists and engineers and to policymakers: it is essential to recognize the increasingly important relationships between scientific and engineering opportunities in this field and to find new ways to coordinate academic, industrial, and governmental institutions to take better advantage of these opportunities. Federal programs have already made substantial progress toward structuring programs to deal coherently with the field of materials science and engineering as a whole. All the institutions working on materials should participate in this trend. A national weakness in synthesis and processing of materials must be remedied: there should be an emphasis on synthesis of new materials, and work on processing should stress science and technology relevant to manufacturing. New facilities and innovation in the development of new instruments for materials research are critical needs.

The United States enjoys a special advantage in analysis, modeling, and numerical simulation; that advantage should be exploited. Materials-based industries in the United States need to revitalize and expand their long-term R&D activities. These are some of the themes of this report; they are described in more detail in the "Conclusions" and "Recommendations" sections below.

CONCLUSIONS

Role of Materials in Industry

Chapter 2 examines the role of materials science and engineering in eight U.S. industries that collectively employ more than 7 million people and have sales in excess of $1.4 trillion, and it summarizes the materials science and engineering needs of each of those eight industries—aerospace, automotive, biomaterials, chemical, electronics, energy, metals, and telecommunications. Several important facts emerged from the industry surveys. Within each industry, several companies were asked to indicate their materials needs; it proved to be possible to describe the needs of those companies in a given industry in generic terms. Furthermore, the lists of generic needs of the various industries had a wide overlap. Finally, industrial materials needs and problems often led scientists and engineers to the frontiers of research in search of solutions. The committee concludes:

- **Materials science and engineering is crucial to the success of industries that are important to the strength of the U.S. economy and U.S. defense.**
- **There is considerable overlap in the generic materials problems of the eight industries studied; solutions to many of these problems lie at the forefront of research in materials science and engineering.**

Two pervasive elements of materials science and engineering that appeared throughout the industry surveys were (1) synthesis and processing and (2) performance of materials. The industry survey participants saw opportunities to improve the effectiveness of all the sectors involved in materials science and engineering. They saw industry as having the principal role in maintaining competitiveness. Accordingly, the committee concludes:

- **The industry surveys revealed a serious weakness in the U.S. research effort in synthesis and processing of materials. There are opportunities for progress in areas ranging from the basic science of synthesis and processing to materials manufacturing that, if seized, will markedly increase U.S. competitiveness.**
- **Increased emphasis on performance, especially as it is affected by pro-**

cessing, is also needed to improve U.S. industrial products for world markets.

• Industry has the major responsibility for maintaining the competitiveness of its products and production operations. Greater emphasis on integration of materials science and engineering with the rest of their business operations is necessary if U.S. firms are to improve their competitive positions in domestic and international competition. Incentives for top-quality people to become involved in production will have to be introduced to achieve such an emphasis. Collaboration with research efforts in universities and government laboratories can enhance the effectiveness of R&D programs too large for any one company. The objective of all of these steps would be renewed emphasis on effective long-range R&D in industry.

Opportunities in Materials Science and Engineering

The practitioners of materials science and engineering have much to say about the challenges and excitement of the field. More than 100 scientists and engineers from many different disciplines and institutions (e.g., universities and industry and government laboratories) participated in this study. Based on evaluation of their contributions, this committee concludes the following:

• The field of materials science and engineering is entering a period of unprecedented intellectual challenge and productivity.

Various properties (or phenomena) that make materials interesting are discussed in Chapter 3. The open intellectual terrain ahead is apparent in each of the materials classes discussed. The structure and properties of materials are understood and are subject to control in ways that were unheard of a decade ago. For example, artificially structured materials can be built up from selected atoms one atomic layer at a time. This reality is deepening and reshaping the concept of what materials science and engineering is. A common element that links the great diversity of work in materials science and engineering is the controlled combining of atoms and molecules in large aggregations in ways that endow the resulting materials with properties that depend not only on the chemical nature of the atomic and molecular constituents but also on their interactions in the bulk of the material and on its surfaces. Calculation of materials properties from first principles is increasingly used by scientists and engineers to understand the origin of properties and to achieve desired characteristics.

These findings are corroborated by other studies of disciplines that play a major role in materials science and engineering [including *Physics Through the 1990s* (1986), *Opportunities in Chemistry* (1985), and *Frontiers in Chemical Engineering* (1988) (National Academy Press)] or by studies of particular areas of materials science and engineering [including *Advanced Processing of Electronic Materials in the United States and Japan* (1986) and *Report on Artificially Structured Materials* (1985) (National Academy Press)]. These studies, as well as the present one, provide examples that suggest that the gap in time between generation of knowledge and application of that knowledge is growing shorter in industries based on materials science and engineering. The processes of basic research, development, and applications engineering are becoming less sequential, separate, and compartmentalized and more concurrent, interactive, and overlapping. Thus the committee concludes:

• **Materials scientists and engineers have a growing ability to tailor materials from the atomic scale upwards to achieve desired functional properties.**

• **In many industries, the span of time between insight and application is shrinking, and these processes are becoming increasingly interactive and iterative. Scientists and engineers must work together more closely in the concurrent development of total materials systems if industries depending on materials are to remain competitive.**

These conclusions surfaced in discussions of research opportunities in structural, electronic, magnetic, photonic, and superconducting materials. From strip casting of metals through the synthesis of new nonlinear optical media in photonic materials, advances in technologies that depend on performance at the cutting edge to remain competitive require the best cooperative contributions of engineering and science.

Emerging Unity and Coherence of the Elements of Materials Science and Engineering

Materials and their applications are diverse, and materials problems involve many science and engineering disciplines. Nonetheless, as discussed in Chapter 4, this committee recognizes an emerging unity and coherence in the field, stemming from the fact that materials scientists and engineers all work on some aspect of materials with the aim of understanding and controlling one or more of the four basic elements of the field. These four elements include:

1. the properties or phenomena that make a material interesting or useful;
2. performance, the measure of usefulness of the material in actual conditions of application;
3. structure and composition, which includes the arrangement of as well as the type of atoms that determine properties and performance; and

4. synthesis and processing, by which the particular arrangements of atoms are achieved.

It is not only these four basic elements—which can be diagramed as a tetrahedron (see Figure 1.10)—but also their relationships that are important.

The scope of materials science and engineering includes not only areas whose utility can be identified today, but also those in which researchers seek a fundamental understanding whose utility may be unforeseen. History has shown time and again that such fundamental understanding leads, often in unexpected ways, to innovations so profound that they transform society. The quantum Hall effect and high-temperature superconductivity are two examples of phenomena involving the collective behavior of electrons in solids that could not have been envisioned a decade ago and whose full implications for our understanding of materials are still evolving. Science in the materials field must include not only those areas whose utility is clear but also basic work that provides fundamental understanding of the nature of materials. Achieving such a fundamental understanding often leads ultimately to important contributions to practical materials problems.

At the engineering end of the spectrum covered by materials science and engineering, there is currently much excitement about the growing ability to exploit the relationships among the four basic elements of the field to develop and produce materials that perform in new or more effective ways. Examples of recent successes extend from the miniaturization of electronic components to steadily improving productivity and quality in the steel industry. Examples of future challenges extend from the practical realization of high-temperature superconductivity to the development of more economical methods of fabricating automotive components from polymers and polymer composites. Thus the committee concludes:

● **Materials science and engineering is emerging as a coherent field.**

● **An effective national materials science and engineering program requires healthy, balanced, and interactive efforts spanning basic science and technology, all materials classes, and the four elements of the field: properties, performance, structure and composition, and synthesis and processing.**

Instrumentation and Modeling

Without advanced instruments, it is impossible to carry out research at the frontiers of science and engineering. In Chapter 4 also, the committee develops the idea that renewed emphasis is needed on research leading to advanced instrumentation and also emphasizes that state-of-the-art instruments are needed to carry out research in the university setting. Such instruments range in size from those at the laboratory-bench scale serving a single investigator to synchrotron radiation facilities serving large numbers of scientists and engineers; they are needed for analysis and for synthesis and processing of materials.

The United States is a leader in the creative use of computers to solve research and engineering problems. Materials science and engineering can be advanced by exploiting this leadership in several areas, from the calculation of electronic-based structures, through simulation of nonequilibrium processes, to real-time monitoring and control of processing. The committee concludes:

● **Progress in the four elements of materials science and engineering can be enhanced through increased R&D on and use of advanced instrumentation ranging from the laboratory-bench scale to major national user facilities, and through increased emphasis on computer modeling and analysis of materials phenomena and properties based on the underlying physical and chemical principles.**

Education

The practitioners in the field come from materials science and engineering departments as well as from various disciplinary backgrounds, including physics, chemistry, and allied engineering fields. Chapter 5 asserts that educating students for careers in materials science and engineering requires a recognition of both the diversity and the coherence of the field.

Many students are immediately employed after receiving a bachelor's degree from a materials-designated department (e.g., a department of materials science and engineering) or from a chemistry, physics, electrical engineering, or other department. Materials science and engineering departments are increasingly emphasizing the four basic elements of the field—synthesis and processing, structure and composition, properties, and performance—to teach a unified approach to all materials at the undergraduate level. The annual production of bachelor's degrees from materials-designated departments is currently about 1000 per year, a figure that has changed little since the 1970s.

Graduate education in materials science and engineering is provided by a diversity of academic departments or divisions, including solid-state physics and solid-state chemistry, polymer physics and polymer chemistry, and engineering, in addition to materials science and engineering and occasionally still other fields such as mathematics or computer science. The annual production of doctorates from these programs is currently just under 700, again about the same as that in the 1970s.

Thus the production of specialists in materials science and engineering has remained essentially constant in the face of greatly increased needs and opportunities in the field. Part of the gap is being filled by an influx of scientists and engineers from other fields (e.g., physicists, chemists, and electrical engineers working on electronic materials). This influx is a continuing source of strength for the field. Part of the gap has not been filled, which has resulted in the current shortage of materials scientists and engineers in universities and industry. The committee concludes:

• **The total number of degrees granted by materials-designated departments plus those granted in solid-state physics and chemistry and in polymer physics and chemistry in the field of materials science and engineering has remained essentially constant for more than 20 years, while opportunities in the field have expanded. If they are implemented, the initiatives recommended in this report will create an additional demand for highly qualified personnel in materials science and engineering.**

• **There is a critical need for curriculum development and teaching materials for educational programs in materials science and engineering to reflect the broadening intellectual foundation of the field and the increased awareness of the importance of synthesis and processing.**

Infrastructure and Modes of Research

Materials science and engineering is practiced at university, industry, and government laboratories. Chapter 6 emphasizes that, although the size of groups working on materials problems varies, most of the effort is carried out on a small scale by individuals or small teams who follow their line of research with modest resources, although some work involves major national facilities. Other work involves larger interdisciplinary teams, and some is carried out by large multidisciplinary groups addressing all four elements of a materials problem (synthesis and processing, structure and composition, properties, and performance). In the long run, there will be a growing need for work on small and large scales to meet the materials challenges of a competitive international marketplace.

Research in materials science and engineering at universities typically is dominated by faculty working independently or in small, sometimes multidisciplinary teams. In contrast, materials science and engineering in industry involves larger, usually multidisciplinary teams. These different approaches will continue to be needed.

The surveys of eight industries referred to above suggest that industry leaders generally consider collaboration with universities desirable and in some cases even essential to address materials problems that must be solved to meet international competition. The committee's survey of materials science and engineering at national laboratories (Chapter 6) suggests that they are also an important resource that is only now beginning to be tapped. Thus the committee concludes:

- **Small-scale research carried out by a principal investigator, sometimes with a small team, is cost-effective and is a major contributor to innovation. The United States has excelled in this mode of research.**
- **Large multidisciplinary teams are an effective mode for addressing industrial materials science and engineering problems.**
- **At the national level, industry, university, and government laboratories have the technical strength to mount major efforts and to exploit break-throughs in the field. All three have been found to be receptive to joint materials science and engineering programs that would be supportive of more rapid commercial development.**

Federal Support for Materials Science and Engineering

Although materials science and engineering is essential for economic and strategic competitiveness, support for materials science and engineering by the federal government has declined in recent years. The data presented in Chapter 6 indicate that during the 11 years from 1976 to 1987, the materials science and engineering budget of the six federal agencies that support most materials science and engineering research declined by 11 percent in constant dollars. The reduction in the nondefense-related portion was even larger— 23 percent. The committee concludes:

- **There is a long-term downward trend in federal support for materials science and engineering that is significantly more pronounced for nonde-fense-related than for defense-related programs.**
- **A strengthened national program in materials science and engineering is necessary to preserve the economic well-being and security of the nation.**

Materials Science and Engineering in Selected Countries

For the last 40 years the United States has led world industry on the strength of its preeminence in science and technology. As Western Europe and Japan have built up their strengths in science and technology, the gap between their status and that of the United States has begun to close. In some areas these nations have caught up with or even overtaken the United States. Chapter 7 points out that the governments of our trading partners have made strong commitments to industrial growth and to coordinated R&D in three areas: biotechnology, computer and information technology, and materials science and engineering. The committee concludes:

- **The governments of the major U.S. commercial trading partners and competitors, including Japan and West Germany, have targeted materials science and engineering as a growth area and as a result have developed strong competence in selected materials science and engineering areas.**

- **These governments have taken a proactive role in deciding which areas of materials science and engineering will be emphasized on the basis of their contribution to enhancing industrial competitiveness.**
- **The various governments use differing mechanisms for achieving national coordination of programs in materials science and engineering, with varying degrees of success.**

RECOMMENDATIONS

The recommendations of this committee are divided into three parts. The first part concerns strengthening the field; the second, maintaining and improving the infrastructure for research in materials science and engineering; and the third, recognizing and developing the unifying trends in the field.

Strengthening Materials Science and Engineering

Finding: Materials science and engineering is a field that is both scientifically and technically exciting and important to mankind through the daily impact of materials on the quality of life. Hence, a strong national effort is justified. The committee's first recommendation is as follows:

- **The national program should include strong efforts in all four basic elements of materials science and engineering—synthesis and processing, structure and composition, properties, and performance. The program should include work that explores the relationships among the four elements and that spans the range from basic science to engineering.**
- **The elements of synthesis and processing as well as performance in relation to processing are currently relatively weak and should be emphasized within this national program.**

Finding: Federal support for materials science and engineering over the past decade shows a downward trend. As a result of the decline in support, the national materials effort is not exploiting new opportunities sufficiently rapidly. In some areas, such as synthesis and processing, there is a shortage of skills and resources. Accordingly, the committee recommends:

- **The federal materials science and engineering program should be restored over the next several years to the levels that prevailed in previous decades in order to exploit the renewed opportunity to make accelerated progress.**

Finding: The general magnitude of the requirements for an adequate national effort in synthesis and processing was discussed with industry representatives. It was apparent that several hundred million dollars would be required to support fully the needs of the electronics and photonics industries alone. Clearly, meeting the needs of all the industries surveyed for this report would require much more support.

Synthesis and processing together form a critically important element of materials science and engineering that has too often been neglected by universities, industry, and government. It is the activity that is responsible for boosting the strength of advanced alloys and composites, for increasing the number of components on integrated circuits, and for producing new superconductors with higher transition temperatures and current-carrying capacities. Work in this area ranges from synthesis of artificially structured materials (with such advanced techniques as molecular beam epitaxy) to engineering of new alloys. Synthesis and processing, which are central to the production of competitive high-quality, low-cost products, lead to new materials with new properties and performance. Work in this area also leads to new and improved production processes with resulting lower costs. The element of synthesis and processing is therefore a crucial determinant of industrial productivity and, ultimately, international competitiveness. The committee recommends:

• **New federal funds should be allocated for support of a national initiative in synthesis and processing. The initiative should provide support for facilities, education, and the development of research personnel. The strengths of universities, industry, and government should be brought into play, and the interactions of these three groups should be directed toward promoting the reduction of materials science and engineering results to commercial practice in the most effective possible manner.**

Finding: Another element of materials science and engineering that needs attention is performance. The properties of materials are put to use by society to achieve desired performance in a device, component, or machine. Some measures of performance include reliability, useful lifetime, speed, energy efficiency, safety, and life cycle costs. Performance is circumscribed by fundamental properties of materials (such as carrier mobility, which influences the switching speed of high-performance transistors, which in turn determines the speed of computers in which such transistors are used). Research to improve performance has received little emphasis in long-range programs, especially in universities, and there has been far too little linkage of this research to the other three elements of materials science and engineering. Some examples of areas representing opportunities for research to improve performance include prediction of the strength and lifetime of complex components and devices, development of improved nondestructive testing techniques, and modeling of systems for optimum material and process selection. The committee recommends:

• **Research on performance (including quality and reliability) should be** increased, especially in relation to processing, but also in relation to the other elements of the field of materials science and engineering.

Finding: Two additional areas of materials science and engineering need greater emphasis: (1) analysis and modeling and (2) instrumentation. In analysis and modeling work, three factors are leading to an explosion of activity, opportunities, and results. The first is the increasing speed, capacity, and accessibility of computers and the concomitant decreasing cost of computing. The second is the growing complexity of materials research and manufacturing. The third is the need in industry to speed the introduction of new designs and new processes into production and to improve production processes and products. Progress in these areas will serve to strengthen fundamental understanding of materials science and engineering and to integrate this understanding with applications. The committee recommends:

• **Increased emphasis should be given to computer-based analysis and modeling in research programs in materials science and engineering.**

Finding: The capability to measure and analyze composition and structure at increasingly smaller levels is surely one of the great engines of progress of modern materials science and engineering. Of equal importance to materials science and engineering progress today is the ability to control structure and composition in new ways and at new levels of precision. Instruments, especially new and sophisticated instruments, will continue to enhance progress in materials science and engineering. The committee notes that the level of support allocated to development of new and unique instruments in universities is small and that U.S. industry is losing its ability to take basic inventions in this area and convert them into business opportunities. The effect of this deterioration in capability is that advanced instrumentation does not diffuse rapidly throughout the academic and industrial research communities. National laboratories, through their large facilities and capabilities in instruments and facility development, may be able to make a unique contribution to this activity. The committee recommends:

• **Government funding agencies should devote a portion of their materials science and engineering program budgets specifically to R&D on and demonstration of new instruments for analysis and synthesis and processing of materials, including instruments that analyze processes in real time.**

Maintaining and Improving the Infrastructure for Research in Materials Science and Engineering

Finding: The field of materials science and engineering is broad. The products of research in this field must meet the exacting standards of intellectual pursuit in an academic setting and of international competition in commerce and defense. The way research is funded and organized in materials science and engineering must reflect this range of goals.

The principal investigator mode of research has made the United States one of the strongest nations in basic research. There is a wealth of good experience with this approach, and the committee has found no evidence to suggest a need to change it. Accordingly, the committee recommends:

- **The U.S. national asset of excellence in the principal investigator mode of research should be preserved and strengthened in the field of materials science and engineering.**

Finding: In recommending preservation of research headed by a principal investigator, the committee recognizes that individuals may join to form small groups to share resources or to attack problems requiring different skills. The committee also recognizes that many principal investigators together may make use of a local resource, for example, a materials laboratory with specialized equipment. On a national level, such investigations can involve cooperative use of a synchrotron light source or a new facility for processing. The committee therefore recommends:

- **A balanced national program of resources, including major national user facilities for materials science and engineering, materials research laboratories, and other regional facilities, should continue to be developed.**

As necessary as an ensemble of principal investigators to carry out research for programs with broad commercial or defense objectives is the involvement of people who understand applications based on new materials or, more frequently, the incremental improvement of existing materials and processes. In order for materials science and engineering to be applied, the coupling between needs and opportunities must be strong. Applied programs need more structure; mutual understanding among those who generate knowledge and those who apply it is essential. This committee has carried out an assessment of the field in this spirit. But materials science and engineering is evolving too rapidly for major decadal surveys such as that done by the National Research Council's Committee on Science and Materials Technology study (COSMAT, *Materials and Man's Needs*, National Academy Press, Washington, D.C., 1975) and the present study to be sufficient in themselves. The committee therefore recommends:

- **Researchers who produce knowledge and those who apply it should continue to work together to identify the needs and opportunities in materials science and engineering, extending the work of this study through periodic reappraisals in selected areas. Such assessments should involve people from university, industry, and government laboratories.**

Finding: The committee has concluded that materials science and engineering is carried out effectively at university and industry laboratories. The committee has observed that government laboratories, including national laboratories under the Department of Energy and the National Institute of Standards and Technology under the Department of Commerce, have considerable strength in people, equipment, and infrastructure to do research in materials science and engineering. Government laboratories have made notable contributions to this field. The strength of all three institutions—universities, industry, and government—should be directed to solving materials problems. Programs developed jointly have several advantages—they define goals, establish needs, identify opportunities, and promote collaboration and communication. The committee recommends:

• **Universities, industry, and government laboratories should develop joint programs in areas of national importance. Government laboratories should play a central role in this effort.**

Recognizing and Developing Unifying Trends in the Field of Materials Science and Engineering

Finding: The broad conclusions of this study are that the field of materials science and engineering encompasses all materials classes; that it spans the full spectrum from basic science to engineering; and that its relation to industrial and other societal needs is strong. The field derives great strength from its relationships to these various entities—the various materials classes, both basic and applied research, and the economic and strategic well-being of the nation. The growing unity of materials science and engineering has implications for universities, industry, and government, as outlined below. The committee recommends:

• **Universities, industry, government, and professional societies should strive to support and to accelerate the unifying trends that already exist in materials science and engineering.**

Universities: Unity in Education

Finding: The subject matter in the majority of materials courses offered in U.S. universities can be taught in a manner that is generic to all materials classes. An adequate curriculum will still contain a few subjects focusing on specific materials (e.g., semiconductors, glasses, metals, and polymers) or on specific functional classes of materials (e.g., optical materials, structural materials, and electronic materials). Such a generic approach to materials science and engineering education depends on exploiting the idea that the field is made up of the elements of properties, performance, structure and composition, and synthesis and processing; this concept provides a unity of subject matter irrespective of materials class or whether a materials problem is examined with the tools of chemistry, physics, or engineering. However, there is a dearth of teaching materials to support such an approach. In some universities, reorganization or new organizational entities may be needed, especially at the graduate level, to achieve a program that will endow materials science and engineering professionals with the breadth and unified view of the field that is now beginning to be expected.

Finding: The most critical resource in any field is well-educated, well-trained personnel. There is a shortage of such individuals in materials science and engineering, especially in the area of synthesis and processing, at all academic levels. The committee anticipates that the increased emphasis on synthesis and processing urged by this study will create an increase in demand for personnel in this area. The committee recommends:

• **Academic programs at the undergraduate level should be oriented to the elements of the field: synthesis and processing, structure and composition, properties, and performance.**

• **At both the undergraduate and the graduate level, increased emphasis should be given to developing new courses and new textbooks that deal generically with all materials. The broadening intellectual foundation of the field and the importance of synthesis and processing should be reflected in these efforts.**

Industry: Collaborating with Universities

Finding: A recurring theme in this study has been the need for stronger university-industry interactions in the field of materials science and engineering. Industry has much to gain from rapid access to advanced basic research activities, to bright future graduates, and to advanced instrumentation. Universities, if they are to remain at the forefront of the field in their teaching and research, must have close and continuing contact with industrial researchers and technologists, and they increasingly will need the financial support of industry. Many ways exist to achieve such a coupling between universities and industry, including joint research activities, joint teaching responsibilities, lifelong education, adjunct professorships, personnel exchanges, scholarship and fellowship support, and support of junior faculty. The committee recommends:

• **Industry and universities should each take the initiative to work together in materials science and engineering with or without government as a partner.**

Government: Bringing the Partners Together

Finding: Given the unifying trends in the field, it is desirable and appropriate that various efforts within relevant agencies have already been consolidated into clearly recognizable units dedicated to materials science and engineering. Renewed efforts to coordinate programs in different federal agencies would be a valuable extension of this accomplishment. Agencies carrying out both extramural and intramural research in materials science and engineering have an opportunity to reinforce their efforts by organizing programs in a way that recognizes the increasingly strong link between the engineering and scientific aspects of the field. A long-range interdisciplinary approach to the entire field is the best approach to capitalizing on the extensive opportunities that it presents. Accomplishing this end is best achieved through formulation and dissemination of broad, long-range goals that go beyond programmatic and disciplinary boundaries. The committee recommends:

- **The government should recognize the essential unity of materials science and engineering in its planning, funding, and coordinating activities.**

Finding: The government plays a leading role in advancing materials science and engineering by supporting basic research at universities and at national laboratories, constructing and operating major user facilities, supporting the enhancement of generic technology in collaboration with industry, performing materials science and engineering germane to the specific missions of each government agency, and developing test methods and reference materials needed for accuracy in characterization of materials.

Finding: The government has additional opportunities to advance materials science and engineering by taking a more active role in the following facilitative functions:

1. *Building consensus.* The government should create mechanisms that will result in the development of consensus among the many sectors that are involved in particular areas of materials science and engineering. Consensus is needed on such topics as evolving research opportunities, the identification of barriers to development that demand broad efforts directed toward their removal, and the understanding and proposing of actions to attack deficiencies in personnel.

2. *Promoting cooperative interactions.* The government should serve as an enabling organization for bringing together various sectors to work on common problems. Objectives could include stimulating the creation of industry consortia, encouraging joint industry-university programs, stimulating joint industry-national laboratory cooperation, and identifying and removing barriers to joint efforts.

3. *Identifying industrial needs.* The government should encourage the

various sectors of industry to identify important materials problems that they anticipate must be solved if they are to improve their competitiveness in the international marketplace. Such problems might include (a) materials needs for products and processes and (b) limitations on analytical capabilities and on availability of data that create barriers to the rapid design, testing, and use of new materials.

4. *Communicating industry needs.* The government should communicate a continuing assessment of the needs of industry that were identified in the eight industry surveys described in Chapter 2 to all members of the materials science and engineering community, including the agencies responsible for supporting materials research.

5. *Balancing federal programs.* The government should establish an annual review process for the federal programs related to materials, including those in research, development, and procurement, to ensure that they are balanced and are responsive to the needs of the nation and the opportunities that are available for accelerating progress.

The committee recommends:

• **The government should assume a more active role in bringing together the various groups involved in materials science and engineering and in enhancing communication, interaction, and coordination among the many sectors affected by materials science and engineering.**

Finding: The National Critical Materials Council (established by P.L. 98-373: Title II—National Critical Materials Act of 1984) has been charged with responsibility for many of the functions listed above.

Finding: Many small businesses that are involved in the materials field can benefit from the availability of new technology and a broader interaction with the larger materials community. State programs that are being established to accomplish these objectives are likely to be more effective than federal ones. The involvement of the state-supported universities, the creation by the states of entities that can effectively experiment with new means of interacting with local businesses, and the willingness of states to invest resources in local enterprises are important and useful developments. The National Institute of Standards and Technology (formerly the National Bureau of Standards), which was created by P.L. 100-418, is envisioned as a possible means of coordination of such state activities.

Finding: The hundreds of laboratories funded by the federal government and sometimes by state governments have many capable personnel and large capital resources that could benefit industry. In particular, the national laboratories funded by the Department of Energy have many scientists and engineers with special talents in materials science and engineering. Reorientation of the missions of the national laboratories toward industrial materials science and engineering interests could have a salutary effect on U.S. industrial competitiveness. (The role of National Institutes of Health laboratories as an asset to the pharmaceutical industry is illustrative.) To be effective in helping industry, federal R&D must be directed intelligently to problems of genuine interest to industry. The federal laboratories, especially the National Institute of Standards and Technology in its new role, could play a useful role in establishing test procedures, setting standards, assembling data collections, and transferring the technology to industry.

• **The committee endorses the goals adopted by the Congress in setting up the National Critical Materials Council, which should work with other agencies to ensure that the government carries out the facilitative functions as well as the more specific tasks identified above.**

• **To accomplish the data collection and analysis that are critical to carrying out these tasks, the committee recommends that the National Critical Materials Council cooperate with other organizations such as the Office of Science and Technology Policy's Committee on Materials, the National Science Foundation, the Department of Energy, the National Institute of Standards and Technology, the National Research Council, and the professional societies.**

1

What Is Materials Science and Engineering?

Materials have been central to the growth, prosperity, security, and quality of life of humans since the beginning of history. Only in the last 25 years, and especially in the last decade, has the intellectual foundation of the field that we call materials science and engineering begun to take shape and to achieve recognition. This has occurred just as the field itself is expanding greatly and contributing significantly to society. Without new materials and their efficient production, our world of modern devices, machines, computers, automobiles, aircraft, communication equipment, and structural products could not exist. Materials scientists and engineers will continue to be at the forefront of these and other areas of science and engineering in the service of society as they achieve new levels of understanding and control of the basic building blocks of materials: atoms, molecules, crystals, and noncrystalline arrays.

MODERN MATERIALS

The fruits of the efforts of materials scientists and materials engineers over years and decades can be illustrated by literally hundreds of examples, and those few given below are but an inescapably arbitrary selection.

• The strength-to-density ratio of structural materials has increased dramatically throughout the industrial age (Figure 1.1). Modern advanced materials are approximately 50 times better than the cast iron of two centuries ago in this important engineering measure. To suspend a 25-ton weight vertically from the end of a cast iron rod would require a rod with a cross

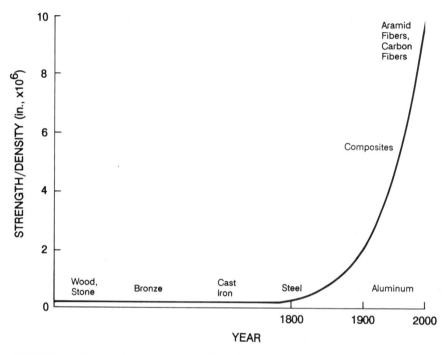

FIGURE 1.1 Progress in materials strength-to-density ratio as a function of time, showing a 50-fold increase in the strength of today's advanced materials compared to that of primitive materials.

section of 1 in. × 1 in. weighing about 4 lb/ft. To suspend the same weight from a modern high-strength polymer fiber would require a fiber with a cross section of 0.3 in. × 0.3 in. weighing about 1 oz/ft. We experience the results of these advances every day, for example, in household appliances that are lighter and more efficient, in eyeglasses that are more comfortable, and in automobiles and airplanes that use less fuel and go faster.

• The efficiency with which heat energy is converted to mechanical or electrical energy in engines and power plants is another engineering measure important to society. This efficiency depends directly on the temperature at which the device can operate well; thus materials that are strong at high temperatures are desired. Superalloys can now operate at temperatures of over 2000°F, and advanced ceramics may push engine operating temperatures to 2500°F (Figure 1.2). The maximum theoretical efficiency of such engines is about 80 percent, whereas the efficiency of conventional engines is limited to about 60 percent. The ultimate result is more efficient production of energy in the forms needed by society, with a concomitant reduction in cost, fuel requirements, and pollution.

• Before about the mid-1930s, the only permanent magnetic materials available were special steels. Modest improvements in the magnetic strength of these materials were made, but significant increases came only with the development of aluminum-nickel-cobalt alloys in the 1940s and 1950s. In the 1960s, the rare earth/cobalt alloys produced the next major jump, and the 1980s saw the development of the neodymium-iron-boron compounds. Today, permanent magnets have magnetic strengths more than 100 times greater than those available at the turn of the century (Figure 1.3). These and other magnetic materials are making possible smaller, more powerful motors and better and smaller sound systems, and they are carrying out many other hidden tasks in modern machines and devices.

• Superconductivity was first discovered in 1911. After some 60 years of effort, researchers developed materials suitable for practical use at temperatures up to 23° above absolute zero (23 K, i.e., 250° below 0°C). Then, beginning in 1986, working with entirely new classes of materials, researchers developed a material with a superconducting transition temperature of 39

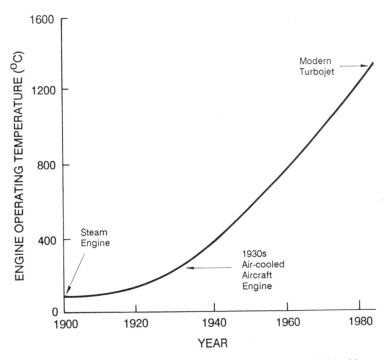

FIGURE 1.2 The steep climb in operating temperatures of engines during this century made possible by modern materials.

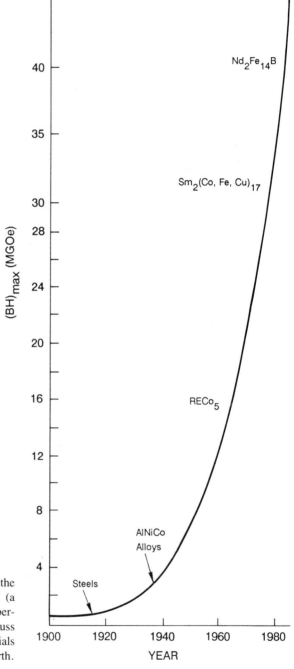

FIGURE 1.3 Progress in the flux-magnetization product (a measure of the strength of a permanent magnet in megagauss oersteds) of magnetic materials over time. Note: RE, rare earth.

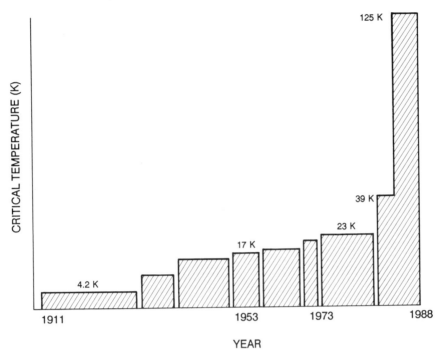

FIGURE 1.4 Progress of critical temperature of the best superconducting material as a function of time.

K. Rapid progress in developing materials with even higher transition temperatures culminated in the present (1988) record of 125 K (Figure 1.4). This discovery not only is of great scientific interest, but it also promises to have a significant practical impact in a wide range of fields. The technical difficulties that prevent the general use of these materials today are precisely those connected with synthesis and processing that contribute the principal challenge to materials science and engineering as a whole.

• Scientists and technicians improved the transparency of silica glass slowly over the centuries from 3000 B.C. to 1966, when work on optical fibers was begun in earnest. Today, these fibers are some 100 orders of magnitude more transparent than they were in 1966 (Figure 1.5). A single glass fiber 0.01 mm in diameter can transmit thousands of telephone conversations—many more by far than can be sent over a conventional cable.

• Even for abrasives and cutting tool materials, it is possible to find significant, often exponential increases in the performance of materials. Figure 1.6 shows, for example, that cutting tool speeds have increased by a factor of 100 since the turn of the century, owing to the development of new materials. The result is far more efficient manufacturing processes that lower the costs of goods we buy.

• In integrated electronic circuits, the number of components per chip has increased at exponential rates since about 1960 (Figure 1.7), making possible the ubiquitous and economical use of the electronic chip we know today. This increase has been achieved partly through steady reductions in line widths through continuing improvements in photolithography (Figure 1.8). Integrated circuits, in turn, have led to computers and electronics that have revolutionized our lives. This achievement is a triumph for both materials scientists and materials engineers, who have mastered the complex interacting relationships between phenomena, materials, and processing.

• Innovations in materials processing have had enormous impacts on the factory floor. In the steel industry, for example, the average worker can now produce 6 times as much steel per hour as he could in the 1920s (Figure 1.9), and the finished steel is of higher quality.

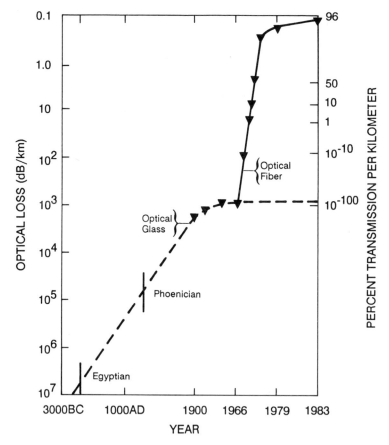

FIGURE 1.5   Historical improvement in glass transparency.

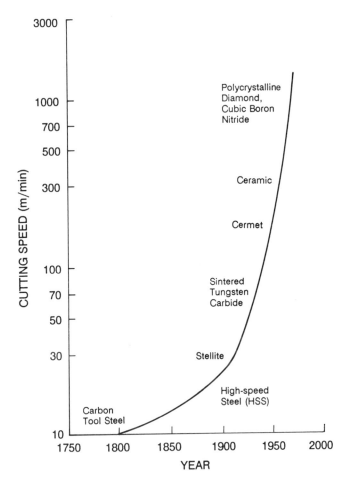

FIGURE 1.6   Trend in speeds of cutting tool materials. (Adapted from M. Tesaki and H. Taniguchi, 1984, High-Speed Cutting Tools: Sintered and Coated, Kogyo Zairyo (Industrial Materials) 32:64–71.)

Materials science and engineering influences our lives each time we buy or use a new device, machine, or structure. Some examples of developments now emerging from our laboratories include the following:

- integrated circuits that will contain as many as a billion components per chip and thus will further the revolution in information technologies that has reshaped modern societies;
- devices that can manipulate and store data optically and thus will contribute to a greatly increased use of optical technology in telecommunications and information storage;

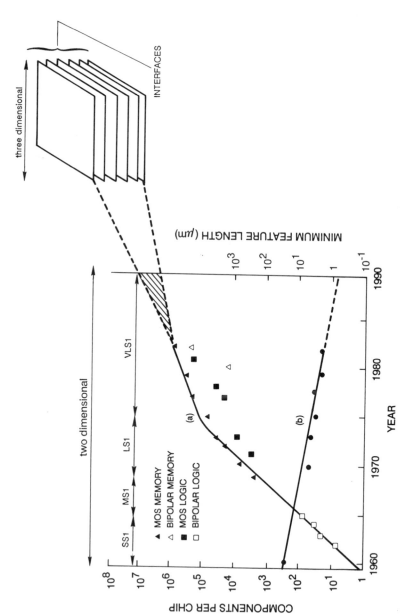

FIGURE 1.7 (a) Exponential growth of the number of components per integrated circuit. (b) Exponential decrease of the minimum feature dimensions. (Courtesy AT&T Bell Laboratories.)

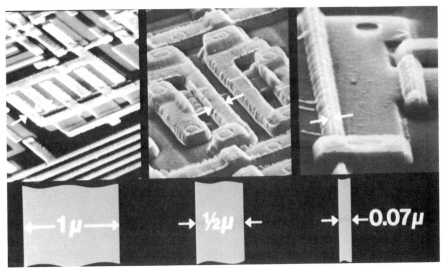

FIGURE 1.8   Micrographs of integrated circuits showing progress in reduction of feature sizes. (Courtesy Praveen Chaudhari, IBM Corporation.)

- materials that can sense, in new ways, their environments of temperature, pressure, and chemistry for control of processes and machines;
- active polymeric materials whose properties, such as color or rigidity, depend on applied electrical or photonic fields, with applications ranging from electronic devices to building materials; and
- biomaterials that can serve as templates for the regrowth of human body parts, such as living tissue or organs.

MATERIALS SCIENCE AND ENGINEERING AS A FIELD

What is the nature of materials science and engineering, a field that so profoundly affects the quality of our lives in so many different ways? The intellectual core and definition of the field stem from a realization concerning the application of all materials: whenever a material is being created, developed, or produced, the properties or phenomena the material exhibits are of central concern. Experience shows that the properties and phenomena associated with a material are intimately related to its composition and structure at all levels, including which atoms are present and how the atoms are arranged in the material, and that this structure is the result of synthesis and processing. The final material must perform a given task and must do so in an economical and societally acceptable manner.

It is these elements—properties, structure and composition, synthesis and processing, and performance and the strong interrelationship among them— that define the field of materials science and engineering. These elements and their relationships are shown schematically in Figure 1.10 in the form of a tetrahedron. In developing new materials, it is difficult to anticipate where seeking knowledge ends and applying it begins. Hence science and engineering are inextricably interwoven in the field of materials science and engineering.

The field of materials science and engineering has evolved along many parallel and intertwined paths associated with academic disciplines, R&D laboratories, and the factory floor. It draws on areas as diverse as quantum mechanics on the one hand and societal needs, including manufacturing, on the other. A proper perspective from which to consider the field requires understanding of the roles of science and engineering and their synergies.

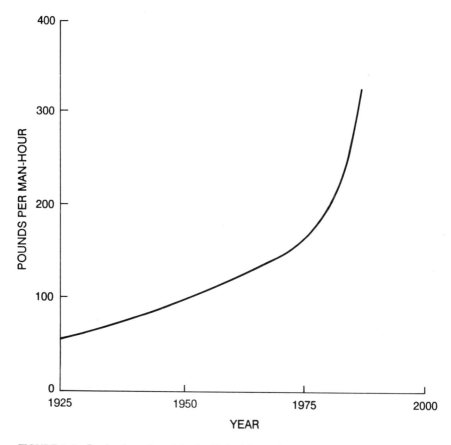

FIGURE 1.9   Production of steel in the United States in pounds per man-hour (1 kg = 2.2 lb).

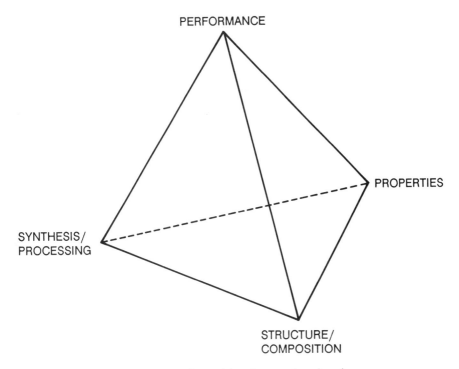

FIGURE 1.10   The four elements of materials science and engineering.

At the science end of its spectrum, materials science and engineering is rooted in the classical disciplines of physics and chemistry. Condensed-matter physicists, solid-state chemists, and synthetic chemists form the bridge between fundamental science and a subset of that science on which modern materials science and engineering rests. These sciences aim to increase knowledge, and especially understanding, of structure, phenomena, behavior, or synthesis. Very often, the stimulation for this group to follow a particular direction of research may come from a technical problem. But the most important advances typically have been made when the research has been placed in a broad context and has been allowed to follow directions whose promise may not have been apparent at the outset.

Earlier in this century, the discovery and understanding of dislocations in crystals revolutionized our understanding of the strength of materials and led to the development of vastly improved structural materials. Understanding of the electronic structures of semiconductors, especially how they are influenced by impurities and in the vicinity of a surface, led to the development of the transistor and, subsequently, to the integrated circuit, which led to the microelectronic revolution.

The deep understanding of the quantum mechanical energy levels of atoms and molecules and of the coupling of electronic motion to light and other forms of radiation made possible the invention of the laser. This discovery, in turn, led to a variety of solid-state lasers, which are used in communication and in information storage.

In three consecutive years—1985, 1986, and 1987—three great advances in fundamental materials science were recognized by the awarding of Nobel Prizes:

1. In 1980, in the course of fundamental studies of electrons moving in semiconductor surfaces, an entirely new and unexpected effect, the quantum Hall effect (associated with almost total absences of electrical resistance), was discovered (Figure 1.11). It should be noted that this research was certainly stimulated in part by the enormous technical interest in electrons near semiconductor surfaces. The experiments were entirely dependent on recent progress in materials science and engineering, which had made possible the preparation of surfaces with extremely well-controlled properties. The full implications of the Hall effect for our understanding of the dynamics of surface electrons are still being developed.

2. In the early 1980s a radically new type of microscope, the scanning tunneling microscope, was developed. It depended on a subtle quantum mechanical effect, the tunneling of electrons *below* the tops of barriers (an event that could not happen under the laws of Newtonian mechanics). This technique led to incredibly accurate information about the positions of individual atoms on surfaces (Figure 1.12). Displacements of the order of 1 percent of the normal interatomic distance can be detected with the scanning tunneling microscope. Instruments of this kind can now be produced very cheaply (for about $25,000) and are responsible for exciting advances in such fields as surface physics, electrochemistry, and biology.

3. Superconductivity was discovered in 1911 in mercury, which lost all electrical resistance below 4.3 K above absolute zero. By the early 1970s, a painstaking materials science effort lasting more than 60 years had led to metallic compounds that remained superconducting at up to 23 K, an average increase of about 0.3 K/yr. No further progress occurred until 1986, when researchers studying entirely different classes of compounds discovered superconductivity at up to 39 K. Since then, based on accumulated knowledge of materials, compounds have been developed that remain superconducting at up to 125 K. Although major materials problems remain to be overcome (e.g., poor ductility and low critical currents), we may well be on the threshold of a technological revolution started by superconduction.

Breakthroughs such as these cannot be predicted or planned, but the environment conducive to their continued occurrence can be. It requires sustained support for and commitment to the basic science that undergirds materials science and engineering.

Collectively, practitioners of materials science and engineering generate

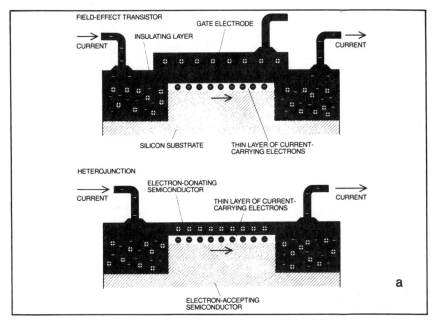

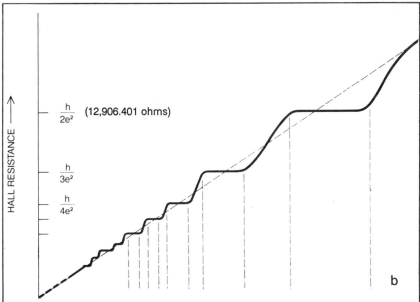

FIGURE 1.11 Quantum Hall effect. (a) Semiconductor devices in which the quantum Hall effect is observed hold current-carrying electrons within a thin layer of semiconducting crystal. (b) Quantum Hall effect appears as plateaus in the Hall resistance of a sample. (Reprinted, by permission, from Bertrand I. Halperin, 1986, The Quantized Hall Effect, Sci. Am. 254:52–60. Copyright © 1986 by Scientific American, Inc. All rights reserved.)

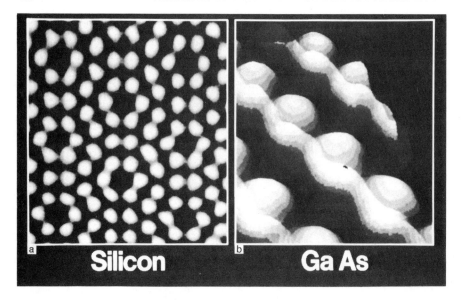

FIGURE 1.12   (a) Scanning tunneling microscope image of silicon surface. (b) Scanning tunneling microscope image of gallium arsenide (GaAs) surface. (Reprinted, by permission, from Praveen Chaudhari. Copyright © 1988 by IBM Corporation.)

and build on the field's scientific base. They understand and exploit the fundamentals of both basic science and engineering, and they translate scientific breakthroughs into forms beneficial to society. Semiconductor devices and integrated circuits would never have been developed if both science and engineering had not been understood; the development of a room-temperature laser required an understanding of semiconducting phenomena and structure (e.g., defects in solids) and of their relationship to processing. Similarly, building and processing of modern composite materials require an understanding of surface science and molecular bonding as well as a strong engineering foundation. Future development of superconducting materials is also essentially dependent on materials science and engineering because such advances require expertise in understanding materials phenomena and their relationship to structure and defects in the structure, to processing (which influences the structure), and to other properties such as brittleness and susceptibility to environmental degradation.

The general approach to solving materials problems for applications came from metallurgists. It was first used at the turn of the century and continued to develop as the relationship between structure, properties, and performance of metals was clearly established. Over time, the important role of processing in controlling structure was realized. The critical role of processing is a central theme of this report. Today, relationships between structure, properties, performance, and processing are understood to apply not only to metals, but also to all classes of materials. Thus modern materials engineering involves exploitation of relationships among the four basic elements of the field—structure and composition, properties, synthesis and processing, and performance (i.e., the elements shown schematically in Figure 1.10), basic science, and industrial and broader societal needs.

Some important materials discoveries have been made by scientists, some by engineers, and still others by craftsmen. Many have been made by teams comprising all three types of individuals. Today, craftsmanship alone, in the absence of modern science and engineering, rarely suffices to bring about a new development in materials. Craftsmanship alone is also increasingly inadequate with respect to processing or production of materials.

A crucial challenge for the future is to find ways of carrying out education, research, and engineering—including production—that encourage the maximum interaction among scientists and engineers, among mathematicians, physicists, chemists, and biologists, and among the four basic elements of materials science and engineering. Supporting such interactions is a difficult task requiring much wisdom, and, realistically, a willingness to make tradeoffs. But progress in accomplishing this task is both possible and essential.

WHO ARE MATERIALS SCIENTISTS AND ENGINEERS?

Materials scientists and engineers study the structure and composition of materials on scales ranging from the electronic and atomic through the microscopic to the macroscopic. They develop new materials, improve traditional materials, and produce materials reliably and economically through synthesis and processing. They seek to understand phenomena and to measure materials properties of all kinds, and they predict and evaluate the performance of real materials as structural or functional elements in engineering systems. This diversity of interests is mirrored in the fields of materials science and engineering practitioners, who come from a broad range of academic departments and disciplines.

There is a growing realization among scientists and engineers that, to develop materials for society, all four elements of materials science and engineering are needed. Even though an individual may identify with a physics, metallurgy, or other department in a university emphasizing a particular aspect of the field, it is implicitly and increasingly recognized that important contributions will come from various disciplines working together.

The interdisciplinary nature of materials science and engineering and its growth as a field have also been recognized in the professional world outside academia. For example, the American Society for Metals, once the largest metallurgical society in the United States, has expanded to become a broad materials society with a new name, ASM International, with a membership numbering about 53,000. The Materials Research Society, established early in 1973, has been one of the fastest growing professional societies in the United States; its current membership includes approximately 8000 individuals. Many industrial and governmental research laboratories have been reorganized into groupings that cut across traditional disciplines and materials areas and have titles such as "materials synthesis," "materials chemistry," and "materials performance." As the science and engineering base of this new field of materials science and engineering develops, so must the process by which its practitioners are educated and the infrastructure and resources with which they approach their task.

SCOPE OF THIS REPORT

This report discusses the vital role that materials science and engineering plays in the development of technology. Chapter 2 summarizes the committee's findings about the impact of materials science and engineering on private and public sector activities that are crucial to U.S. economic and strategic well-being. Opportunities for research are discussed from two perspectives: Chapter 3 describes needs for new materials and for novel methods of processing in terms of the functional roles of materials; Chapter 4 describes research opportunities in the context of the four basic elements of materials science and engineering, thus emphasizing the intellectual coherence of the field while also stressing the essential connection between basic research and progress in developing materials. Educational challenges posed by the national need to encourage such progress and to ensure an adequate supply of well-trained materials researchers are considered in Chapter 5, which briefly assesses resources available for educating materials scientists and engineers at various levels of the U.S. educational system and also emphasizes the significance of the field's multidisciplinary aspect. Chapter 6 presents the committee's findings about funding and facilities currently available—as well as those needed in the future—to support the research efforts of materials scientists and engineers who work at the perennially shifting boundary between gathering knowledge and applying it. Finally, to examine from a broader perspective its assessment that materials science and engineering is vital to the future development of U.S. technology, the committee also examined how a number of nations view materials science and engineering and its role in their development. The international perspective is presented in Chapter 7. The committee is convinced that its findings and recommendations, if implemented, will strengthen the field of materials science and engineering and, in so doing, will contribute immeasurably and in unanticipated ways to meeting U.S. needs for economic and strategic security as well as the future needs of mankind.

2

Materials Science and Engineering and National Economic and Strategic Security

This chapter examines the impact of materials science and engineering on U.S. society. The committee evaluated the impact of materials science and engineering by surveying its role in industries considered important for commerce and defense and then looking briefly at needs of the public sector, particularly of governmental units whose missions involve defense, energy, transportation, and space.

SIGNIFICANCE OF MATERIALS SCIENCE AND ENGINEERING IN INDUSTRY

Eight industries that represent different aspects of the use of materials were chosen to be surveyed, including the aerospace, automotive, biomaterials, chemical, electronics, energy, metals, and telecommunications industries. The scope of each of the eight surveys is shown in Table 2.1.

The surveys of the eight industries were carried out by people with senior management and technical responsibilities in their respective industries. Hence the results of the surveys are particularly important in two respects. First, they represent a sample of industry views regarding materials science and engineering and its impact. Second, they represent technical management views on how materials science and engineering should be structured by policymakers to fully exploit the opportunities that lie ahead. The results of the surveys show that materials science and engineering is viewed as vital by all eight industries. The idea also emerged that it is important for government to play a leadership role in helping to identify research areas of

TABLE 2.1   Industries Surveyed in This Study

Industry	Scope of Survey
Aerospace	Airframe and engine materials (not electronics)
Automotive	Primarily automobiles
Biomaterials	Primarily materials used in contact with human body tissue
Chemical	Traditional chemicals, polymers, advanced ceramics
Electronics	Materials for computers, commercial and consumer electronics
Energy	Electricity, coal, oil, natural gas, nuclear, solar, geothermal
Metals	Production and forming of primary metals
Telecommunications	Materials for telephone and data transfer equipment

national importance, so that materials science and engineering can be more fully exploited.

The eight industries collectively employed 7 million people and had sales of $1.4 trillion in 1987 (Table 2.2). In addition, they were critical to many millions of jobs and to huge sales in ancillary manufacturing industries, for example, in the manufacture of materials for electronic applications that drive the computer hardware industry.

TABLE 2.2   Economic Impact of the Eight Industries

Industry	1987 Employment[a] (thousands)	1987 Sales ($ billion)
Aerospace	835	105.6
Automotive	963	222.7
Biomaterials	—	>50
Chemical	1004	195.2
Electronics	1394	155.4
Energy	1229	375.8
Metals	629 (1230)[b]	98.9
Telecommunications	1007	146.0

[a]The statistics are taken from the *U.S. Industrial Outlook 1989*, published by the Department of Commerce, *International Trade Administration*, Washington, D.C.

[b]The 1980 to 1985 average based on a broader definition of the metals and mining industry used in *Employment Prospects for 1995*, Bulletin 2197 published by the Bureau of Labor Statistics, Washington, D.C. (1984).

The recent economic performance of the eight industries has varied widely. The U.S. metals industry, which is still very large, has declined significantly overall in employment and sales over the last decade. Nonetheless, in 1988 much of the industry was operating at capacity, and exports were once again on the increase. At the other extreme, the biomaterials industry, which started from a very small base, is in a period of very rapid growth.

Table 2.3 shows the international trade balances for seven of the eight industries surveyed. The aerospace and chemical industries are healthy exporters and contribute substantially to the U.S. position in international trade. Although the trade balance for the chemical industry has declined somewhat as production of petrochemicals has grown in the Middle East and manufacture of synthetic apparel fibers has shifted to the Far East, this negative trend seems to have slowed recently. As is well known, imports of automobiles and petroleum have had an extremely negative effect on the U.S. balance of payments. A particularly worrisome trend is the decline in the trade balance for high-technology industries such as electronics and telecommunications. The biomaterials industry, in which the United States has a strong position, is omitted from the table because the industry is comparatively small.

Of the eight industries surveyed, two are primarily producers of materials. The metals industry has a well-defined traditional role as a producer of bulk and formed metals. However, the chemical industry, which historically has been a supplier of bulk chemicals and polymers, is undergoing rapid change. American chemical companies are diversifying into biotechnology, materials for the electronics industry, ceramics, and specialty metals (such as amorphous metals prepared by rapid solidification techniques). In fact, they are becoming broad-spectrum producers of materials, with an emphasis on high-value products. To some extent, the growth of the biomaterials industry is occurring under the wing of the chemical industry.

TABLE 2.3   International Trade Balances for Seven Selected Industries (billions of dollars)

Industries	1982	1984	1985	1986	1987
Aerospace	+ 11.1	+ 10.2	+ 12.3	+ 11.7	+ 15.1
Automotive	− 10.4	− 20.7	− 26.5	− 35.8	− 42.4
Chemical	+ 12.4	+ 10.7	+ 8.5	+ 8.5	+ 9.3
Electronics	+ 6.7	+ 2.5	+ 2.6	+ 0.6	− 0.1
Energy	− 53.3	− 52.7	− 44.2	− 30.8	− 38.3
Metals	− 9.5	− 12.9	− 11.6	− 9.6	− 10.8
Telecommunications	+ 0.2	− 1.0	− 1.2	− 1.3	− 1.7

SOURCE: Data are abstracted from *U.S. Industrial Outlook 1989*, published by the International Trade Administration, Department of Commerce, Washington, D.C.

The metals, chemical, and biomaterials industries are also consumers of materials. The processing equipment for metals and chemicals often requires materials resistant to high temperatures and to corrosive environments. New materials with outstanding resistance to heat and corrosion can be critical to the success of a new process technology. In the chemical industry, selectively permeable polymeric membranes are beginning to have an impact in new separation processes based on dialysis and reverse osmosis.

In industries that might be considered primarily consumers of materials, the roles of materials vary widely. The aerospace, automotive, and energy industries are most concerned with structural materials, whereas the electronics and telecommunications industries emphasize development of materials that have an active function. Biomaterials generally serve both structural and functional roles. The more rapidly evolving segments of these industries are active in the development of new materials such as composites.

The aerospace industry and, to a lesser extent, the automotive industry have a major interest in reducing the weight of their structural materials to increase fuel economy and performance. Although approximately half the cost of a modern aircraft lies in its electronic gear, reducing the weight of the airframe can significantly reduce the cost of its operation. There is a similar interest in high-temperature materials for highly efficient aircraft engines that will also decrease fuel consumption. Because of the large economic impact of improvements in these areas, the aerospace industry has become a major developer of advanced materials.

The energy industry has many different segments with different materials needs. On the one hand, coal, petroleum, and natural gas production has only marginal, incremental needs for new materials. On the other hand, the fossil and nuclear power and solar energy segments can benefit greatly from materials with improved performance. New developments such as high-temperature superconductivity may have a profound influence on the production, transmission, and use of electricity.

Because improvements in performance in the electronics and telecommunications industries are closely tied to improved electronic and optical properties of materials, these industries play a dynamic role in developing new materials and processes. The link to materials is especially close because fabrication of a semiconductor device, for example, often involves synthesis of functional materials in situ.

The biomaterials industry is unique in that its products must be compatible with body tissue, and new materials must be approved for use by the Food and Drug Administration. These requirements present special challenges for materials developers.

Some of the generic materials needs of the eight industries are summarized in Table 2.4. These needs, in turn, represent opportunities to improve the economic performance of the industries, as discussed below.

TABLE 2.4 Materials Needs of the Eight Industries

Desired Characteristic	Industry							
	Aero.	Auto.	Bio.	Chem.	Elec.	Energy	Metals	Telecom.
Light/strong	✓	✓	✓					
High temperature resistance	✓			✓		✓	✓	
Corrosion resistance	✓	✓	✓	✓		✓	✓	
Rapid switching					✓	✓		✓
Efficient processing	✓	✓	✓	✓	✓	✓	✓	✓
Near-net-shape forming	✓	✓	✓	✓	✓	✓	✓	✓
Material recycling		✓		✓			✓	
Prediction of service life	✓	✓	✓	✓	✓	✓	✓	✓
Prediction of physical properties	✓	✓	✓	✓	✓	✓	✓	✓
Materials data bases	✓	✓	✓	✓	✓	✓	✓	✓

These findings are consistent with the results of an international survey, discussed in Chapter 7, that clearly shows that many of the major trading partners of the United States have targeted research in materials science and engineering, along with biotechnology and computer and information technology, as one of three principal areas for special growth. They have also targeted specific areas within materials science and engineering for development in their nations.

Aerospace Industry

Scope of the Industry

The aerospace industry is large and dynamic. In 1987, it employed 835,000 workers (a figure that doubles when supplier companies are included) and had sales of $105.6 billion (see Table 2.2). The industry has had a consistently positive balance in international trade, including $15.1 billion in 1987 (see Table 2.3). Despite the traditional technological leadership of the U.S. aerospace industry, however, extremely stiff foreign competition has developed. Beyond its role in the civilian economy, the industry is critical to the national defense.

The survey of the aerospace industry covered both military and civilian airframe and engine production as well as materials needs for spacecraft. Electronic materials for aircraft applications were excluded from this survey, because they were included in the electronics industry survey.

Role of Materials in the Aerospace Industry

The aerospace industry is both a user and a developer of high-performance materials. Aerospace systems push structural materials capabilities to their limits.

The industry must constantly revise its practices to ensure continued system reliability and safety and structural integrity. Advanced materials such as composites are used extensively in military aircraft, helicopters, and business planes. In large civilian transport aircraft, the introduction of such materials is much slower (Table 2.5); they appear primarily in secondary structures. Broader use of composite materials will require large changes in design and in manufacturing plants.

Advances in turbine airfoil materials are illustrated in Figure 2.1. The volume of materials consumed by the industry is not large (e.g., about 80,000 tons/year for large commercial aircraft), but the value is extraordinary. The cost of a commercial airframe is approximately the value of its weight in silver. The cost of a spacecraft approximates the value of its weight in gold. Because of these economic factors, substantial costs can be tolerated for materials that possess the desired combination of properties.

Needs and Opportunities

Some principal determinants in the selection of materials for the aerospace industry are life cycle cost, strength-to-weight ratio, fatigue life, fracture toughness, survivability, and reliability. Additional considerations for spacecraft include high specific stiffness and strength, a low coefficient of thermal expansion, and durability in a space environment.

The payoff for successful materials development can be large. In a shuttlelike orbiter, for example, replacement of conventional aluminum airframes with currently unobtainable aluminum/silicon carbide or magnesium-graphite composites would yield a severalfold increase in payload capability. Similarly, a major reduction in airframe weight could lead to an increase in fuel efficiency that would make an aircraft attractive to commercial airlines. Lifetime sales for a successful new generation of commercial transport planes could be expected to amount to about $45 billion.

Conventional materials (e.g., metals, alloys, ceramics, and polymer composites) are approaching developmental limits in terms of properties for aerospace applications. This limit is based on fatigue (or service life) criteria

TABLE 2.5   Use of Materials in Civil Transport Airframes

Material	Percentage of Structural Weight by Year	
	1987	2000 (projected)
Aluminum	71	55.5
Composites	7.2	24.8

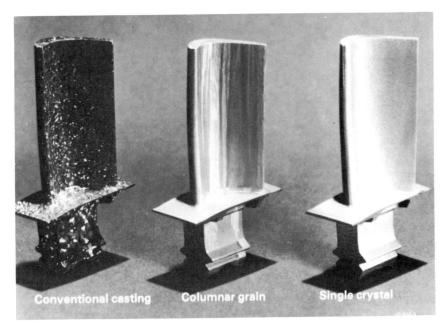

FIGURE 2.1   Progress in casting techniques for turbine blades. Standard methods produce a polycrystalline blade (left). With directional solidification, the crystalline structure is oriented in the direction of the stresses encountered in operation, imparting greater strength and creep resistance to the blade (center). The blade on the right is a single crystal, which is even stronger. (Reprinted, by permission, from Bernard H. Kear, 1986, Advanced Metals, Sci. Am. 255:159–167. Copyright © 1986 by Pratt & Whitney Aircraft.)

as well as on strength at elevated temperatures. Innovative research and engineering are needed to provide high-strength and/or heat-resistant ultra-light structures for use in advanced subsonic, supersonic, and transatmospheric aircraft. Opportunities exist to develop and use composites of all types, including new alloys and intermetallics as well as multilithic composites such as metal matrix, ultrafine metal-metal, cermets, ceramic-ceramic thermoplastic, thermoset-thermoplastic, and molecular polymeric types. Designs must be modified to accommodate these materials in a cost-effective way.

New metallic alloys such as aluminum-lithium, intermetallics like titanium aluminides, and high-temperature alloys derived from rapid solidification technology (e.g., aluminum-iron-vanadium-silicon) offer promising avenues for research on materials with good strength-to-weight ratios. Metal matrix composites are an especially ripe area for development, with applications in layered metal structures and especially in ceramic-reinforced metals.

Ceramics appear to be very attractive for high-temperature applications such as radiant burner tubes and leading edge structures for wings. With respect to polymeric materials, carbon-carbon composites retain high strength at high temperatures in hostile atmospheres. A wide range of properties for use at lower temperatures is accessible from combinations of polymeric binders and reinforcing materials. So-called molecular composites offer many opportunities. Stiff, strong, chemically compatible reinforcing materials continue to find increasing use in high-performance composites.

This description of materials and the possible opportunities they offer strongly emphasizes composite materials and suggests the consequent requirement for major advances in our understanding of the chemistry and physics at the interfaces between dissimilar materials. In addition to this understanding at the molecular and microstructural level, major advances in processing and fabrication technology are required. The simultaneous development of materials, processing, and fabrication is essential if the new technology is to be used in an efficient and timely fashion.

Cost-effective, high-quality processing technology is essential. Real-time, on-line process control systems, computer modeling, and advanced sensor development must complement fundamental materials science. As in other industries, materials development requires a systems approach encompassing materials preparation, processing, fabrication, quality assurance, and in-service monitoring. The research on processing must be coordinated among the aerospace users and developers and the materials suppliers who will ultimately produce the materials.

Automotive Industry

Scope of the Industry

The production of automotive products is one of the largest components of the U.S. economy. The industry had sales of $223 billion in 1987 and employed about 1 million persons directly, as well as many others in supporting businesses. As noted below, the production of motor vehicles makes the industry one of the largest consumers of materials.

The automotive industry is a major cause of the current U.S. trade deficit. In 1986, automobile imports cost $46.5 billion and resulted in a net trade imbalance of -35.8 billion. Clearly, if these imported vehicles had been made in the United States or if there had been offsetting exports, the U.S. economy and the U.S. automotive and metals industries would have been much healthier. The international competitiveness of the automotive industry is crucial to the whole U.S. materials industry.

The survey done for this study covered materials requirements for the production of cars, trucks, and buses in the United States.

TABLE 2.6 Use of Materials in the
Automotive Industry in 1984

Material	Percentage of Total U.S. Consumption
Steel	17.5
Aluminum	16.0
Copper	9.7
Lead	59.0
Platinum	46.9
Zinc	26.0
Synthetic rubber	55.1
Natural rubber	76.6

Role of Materials in the Automotive Industry

The production of motor vehicles consumes 60 million tons of metals, polymers, ceramics, and glasses per year. Table 2.6 lists the automotive industry's share of total U.S. consumption of basic materials in 1984.

As a consumer product, the automobile must be made from materials that are cheap and easy to process, and it must have long life and high reliability under extremely adverse conditions. In recent years, there has been additional pressure to make vehicles lighter in the interest of fuel economy. This new requirement has led to extensive substitution of aluminum and plastics for steel and heavy metals and has led to extensive changes both in vehicle design and in manufacturing processes.

Needs and Opportunities

Improvements in materials can have a large impact on the economic health of the automotive industry. One important need is the development of complete materials systems that take into account the cost of production and fabrication of a material along with specific design criteria.

The survey identified four major needs that may be considered driving forces for automotive technology. These basic requirements and their implications for materials science and engineering are listed in order of priority in Table 2.7. A fifth requirement, reusability of materials, is significant because it also emerged in the metals industry survey as a high priority.

Research Opportunities

The survey identified R&D needs for 11 major classes of materials used in automobiles: sheet steels, specialty steels, structural plastics and composites, nonstructural plastics and composites, elastomers, paint, nonferrous

TABLE 2.7 Materials Needs for the Automotive Industry

Generic Drivers of Automotive Technology	Implications for Materials R&D
1. Need for reliability	Emphasis on production of materials with minimal variation in properties and dimensions
	Emphasis on processes that can be used to convert materials to components with minimal variation in size and shape
	Development of processes simultaneously with materials
	Development of sensors and control systems of materials production and fabrication processes
	Development of materials data bases including variation of properties
2. Need for low cost	Emphasis on low-cost (including energy) and in particular low-cost materials production systems
	Emphasis on development of materials that can be processed at low cost and emphasis on low-labor and high-throughput and low-waste process development
	Development processes simultaneously with materials
	Research on existing as well as "new" materials systems
3. Need for functional improvement	Emphasis on weight-reducing materials
	Materials with improved conductivity and catalysis
	Development of functional characteristics and design studies simultaneously with materials and process development (a "materials" systems approach)
	Materials with improved sound absorption, toughness, transparency, dent resistance, etc.
4. Need for durability	Research on durability-related failure mechanisms (wear, fatigue, and corrosion)
	Research on methods of predicting durability
5. Need for reusability	Research on technical and economic factors in recycling

wrought metals, castings, ceramics, tool and die materials, and metal matrix composites. Plate 1 illustrates where some of these materials are used in an automobile.

Research is needed to increase understanding and accessibility of materials properties in all classes of materials, polymers and composites as well as metals. Predictive models for both properties and processing can have a significant impact. Intense R&D effort directed to composites could lead to a clear U.S. competitive advantage. The United States currently leads in this area but must work hard to maintain its lead. Ceramics have important roles as tool and die materials as well as in engine components such as turbochargers. Sheet steels, castings, plastics, and structural composites—particularly their processing—offer the largest potential opportunities to improve automotive technology.

Biomaterials Industry

Scope of the Industry

The biomaterials industry encompasses the design, fabrication, and manufacture of materials for the health and life sciences fields. It has been estimated that the industry has sales of over $50 billion annually and is growing at a rate of 13 percent per year. The industry's products include disposable hospital supplies, artificial organs, personal care products, diagnostic devices, drug-delivery systems, and separation systems for biotechnology. Its participants range from operating divisions of Fortune 500 companies to small, family-operated businesses. The industry has a positive balance of payments in international trade.

The United States has had a leading role in the development of the biomaterials industry and is a strong exporter, as exemplified by its dominant market share in the European Economic Community countries. To counter U.S. dominance, Japan, South Korea, Italy, and Sweden are moving aggressively to build their biomaterials and biodevice industries.

The industry can be divided into market segments as follows:

- Artificial organs
- Biosensors (diagnostic devices)
- Biotechnology
- Cardiovascular and blood products
- Drug delivery
- Equipment and devices
- Maxillofacial prostheses, materials for plastic surgery
- Ophthalmology
- Orthopedics
- Packaging (including hospital supplies and consumer products)
- Wound management

Role of Materials in the Biomaterials Industry

The biomaterials industry develops, produces, and uses materials of extraordinary diversity. Currently used materials include synthetic polymers (degradable and nondegradable), water-soluble polymers, biopolymers, metals, ceramics, glasses, glass-ceramics, carbons, and biologically derived materials. Figure 2.2 shows artificial skin composed of a layer of silicone rubber and a layer of modified collagen.

In specialized applications such as artificial organs, ophthalmic lenses, and specialty catheters, very high costs can be tolerated for materials whose physical and biological properties are satisfactory. However, consumable items such as hospital supplies and personal care products are price sensitive, and the cost of materials becomes a significant factor. Figure 2.3 shows a drug-delivery system that employs a microporous membrane to introduce drugs through the skin.

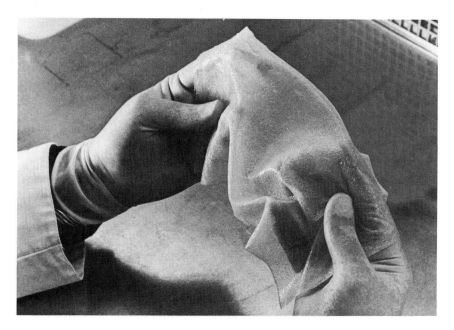

FIGURE 2.2   Artificial skin has been used successfully to treat more than 100 patients, all victims of severe burns. Clinical trials in 11 hospitals have established the potential of this membrane as a substitute for the conventional autograft treatment. The outer layer of this bilayer membrane is silicone rubber and is removed and disposed of 2 weeks after grafting. The inner layer, which makes contact with the wound, is a highly porous layer of modified collagen that itself is degraded about 10 days after contacting the wound. Before degradation is complete, the collagen layer acts as a biological template that induces partial regeneration of new dermis. (Reprinted, by permission, from the Massachusetts Institute of Technology News Office.)

Needs and Opportunities

Eventually there will be a need for biomaterials that duplicate the physical and biological properties of all native tissues in the body. Some examples are given in Table 2.8. Requirements for these materials include the following:

- Nonthrombogenic surfaces (surfaces that do not promote the formation of clots)
- Hydrolytic stability
- High purity

- Reproducible quality
- Stability to sterilization
- Biocompatibility
- Bioreabsorbability

Research Opportunities and Issues

Development of complete materials systems is needed for biomaterials in the areas of artificial organs, extracorporeal blood treatment, drug delivery, and biotechnology. Such systems must include production of the material, processing, fabrication into a device, and testing of the device. These tightly coupled processes must preserve desired biological properties. Life cycle

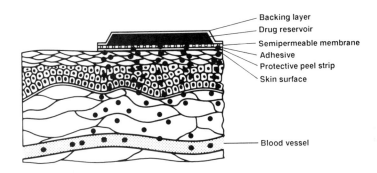

- Backing layer
- Drug reservoir
- Semipermeable membrane
- Adhesive
- Protective peel strip
- Skin surface

- Blood vessel

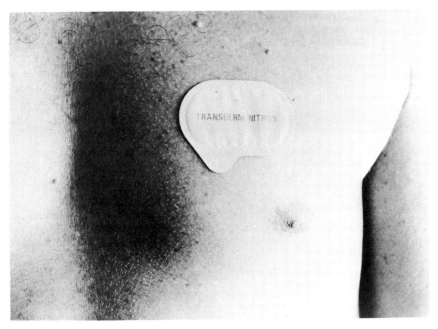

FIGURE 2.3 Transdermal drug delivery system. (Reprinted, by permission, from Ciba-Geigy Corporation. Copyright © 1989 by Ciba-Geigy Corporation.)

TABLE 2.8   Needed Biomaterials

End Use Application	Material Need
Burn/wound coverings	Grafts for epithelium cell regeneration
	Release of antibacterials
Cardiovascular implants	Thromboresistant surfaces
	Small-diameter vascular grafts (less than 4-mm diameter)
Catheters:	Thromboresistant skin
cardiovascular,	Infection-resistant surfaces
urinary	Nondenuding (able to slip over epithelial tissue without adhering and stripping)
Controlled release	Bioadhesives
	Bioerodable polymers
	Protein delivery
Diabetes	Hybrid artificial organs
Extracorporeal blood	Immobilized chemotherapeutic treatment agents and enzymes for chemotherapy and detoxification
Neural repair	Polymers that induce nerve regeneration
Ophthalmologic	Artificial corneas
	Vitreous implants
Orthopedic	Fiber composites
	Resorbable polymers
Soft tissue reconstruction	Resorbable polymers with concurrent release of bioactive agents
Wound closure	Tissue adhesives

engineering must be part of the design. Several generations of devices envisioned include the following:

- Devices incorporating biochemical and pharmacological activity
- Hybrid devices containing biological tissue in a synthetic matrix
- Cultured organs and tissues

Chemical Industry

Scope of the Industry

The American chemical industry is a broad-based supplier of materials, not only chemicals and polymers but also biomaterials, electronic and optical materials, ceramics, and specialty metals. The established chemical and polymer industry contributed over $195 billion in sales to the U.S. economy in 1987 and employed over 1 million people. In addition, sales of advanced ceramics amounted to about $4 billion to $5 billion. [In *U.S. Industrial Outlook 1987* (published by the International Trade Administration, Department of Commerce, Washington, D.C.), advanced ceramics are exemplified by high-performance ceramics for jet engines, machine tools, electronic packaging, and solar energy devices.]

The U.S. chemical industry has consistently had a positive balance of payments in international trade. However, this favorable position has shrunk in recent years from $12.4 billion in 1982 to $9.3 billion in 1987. Certain major sections of the U.S. chemical industry, such as the manufacture of apparel fibers, are becoming less competitive. The chemical industry is undergoing major changes, moving away from commodity materials toward more highly engineered products such as optical discs and AIDS diagnostic systems. People speak of a chemical industry revolution as large as the one brought on by the advent of synthetic polymers. The impact of one of these, polymer composites, suggests the importance of these new ventures (Figure 2.4). According to industry analysts quoted in *Chemical and Engineering News* (March 16, 1987), sales of polymer "composite to certain specialty markets will grow from $2.5 billion in 1986 to $3.7 billion in 1991. . ." and will be ". . . a major growth market with an annual value of $10 billion by the late 1990s."

The survey of the chemical industry stressed the materials (including

FIGURE 2.4 Filament-wound pressure vessel. (Reprinted, by permission, from E. I. du Pont de Nemours & Co., Inc. Copyright © 1989 by E. I. du Pont de Nemours & Co., Inc.)

ceramics) that the chemical industry supplies to other industries and treated to a lesser extent the materials needs of the chemical industry.

Role of Materials in the Chemical Industry

The major role of the chemical industry is to supply materials. The industry processes raw materials to form new, higher-value materials that are sold to other industries. The chemical processing industry has continuing needs for strong, noncorroding materials that can withstand extreme conditions and for materials that facilitate chemical processes, such as catalysts and electrode materials. All chemical processes incorporate some materials separations. Major breakthroughs are being made in the science of the materials that effect these separations—zeolites, ion-exchange resins, affinity chromatography materials, membranes, supercritical fluids, and others.

Needs and Opportunities

The materials being developed and produced by the chemical industry include polymers, ceramics, fluids, composites, and single crystals. New materials under development include polymers for automotive body panels, extended-chain polymers (e.g., polymers spun from liquid crystals), thermoplastic matrices for advanced composites, electrically conductive polymers, polymers with nonlinear optical properties, and biocompatible polymers. The use of metal-forming techniques such as forging and plasma spraying affords additional opportunities.

Recently, the chemical industry has entered the field of ceramics research in the belief that high-performance ceramics will be made and processed as chemicals, rather than by direct processing of minerals. Active areas of research include structural ceramics, ceramic fibers and whiskers for composites, low-temperature processing, chemical and polymeric precursors, ceramics for electronic applications with improved dielectric properties, and the exciting studies of diamond films.

Little systematic research has been done on fluids, but they have myriad uses as refrigerants, solvents, and hydraulic fluids. Oxidative and mechanical stability are two long-standing requirements for products in these applications; in addition, innocuousness of effects on stratospheric ozone is becoming a major criterion for new products.

Composites account for an extremely active area of R&D. Polymer, ceramic, and metal matrices with dispersed fibers, whiskers, or other particulates are all being studied. In many cases the composites have properties superior to those of their constituents. Most of the work to date has been empirical, but understanding and control of the chemical changes that occur in the creation of a composite will be critical to the development of this field. The opportunity for a positive economic impact is enormous.

Crystalline materials are critical to advances in the electronics and optics industries. Epitaxial growth processes are key to the creation of artificially layered structures, which in turn are opening exciting new areas in electronics and optics. Purity of materials, for which requirements are already demanding, must be further improved for applications now envisioned.

Electronics Industry

Scope of the Industry

The electronics industry, one of the most dynamic industries in the global economy, is a leader in the use of new materials. The leverage of materials science and engineering on the electronics industry through added value is extraordinary. The 1985 worldwide electronic materials market of $2.5 billion was responsible for an equipment market of roughly $400 billion. In the United States in 1987, the electronics industry (excluding instrumentation and telecommunications equipment) contributed about $155 billion in revenue to the economy. The broad markets were distributed as shown in Table 2.9.

There is a consensus that the U.S. electronics industry is losing ground to competition from the Far East, primarily because of manufacturing and marketing leadership abroad. Although the United States still leads in basic materials science and characterization, our growing dependence on foreign suppliers for critical components and manufacturing equipment is a threat to national security.

Recent trends in the semiconductor industry illustrate the economic penalty associated with deficiencies in our ability to process and manufacture semiconductors cheaply and well. The U.S. share of the global semiconductor market fell from 61.8 percent in 1980 to 43.3 percent in 1986. Retaining a 61.8 percent share of the 1986 market of $26.4 billion would have added $4.9 billion in revenue for the U.S. industry. More importantly, a loss of $1 in semiconductor sales generally results in the loss of $10 in sales of

TABLE 2.9   Electronics Industry Employment and Sales in 1987

Sector	Employment (thousands)	Sales ($ billion)	Trade Balance ($ billion)
Electronic components	520	47	− 2.4
Computing equipment	326	54	+ 2.8
Radio and TV	548	54	− 0.5
Total	1394	155	− 0.1

electronic systems. Thus our loss of competitiveness in semiconductors may have cost the U.S. electronics industry about $50 billion in 1986. Given that semiconductor industry sales are projected to be $60 billion in 1992, the future leverage is even greater. More efficient coupling of U.S. materials science and engineering research with electronics manufacturing is vital for the health of this country's economy.

Role of Materials in the Electronics Industry

New, highly engineered materials are vital to the progress of the electronics industry. Strong interrelationships between materials science, device design, and fabrication process chemistry determine the performance of an electronic device. Multidisciplinary approaches are needed for rapid development and transition to production. The major focus of this survey was materials for microelectronics, but magnetic and display materials were also considered.

Semiconductor materials can be used to illustrate materials development in the electronics industry. Single-crystal silicon (150-mm diameter) is grown from the melt by a highly automated process. Oxygen levels for gathering impurities and slice strengthening are controlled by computer. Devices made from wafers cut from the crystal are now made mostly by epitaxial growth. A variety of processing techniques (e.g., photolithography, plasma chemistry, and epitaxial film deposition) are used to form the devices (Plate 2). Similar processes and technologies are also used with III-V semiconductors, but they are not as advanced. The realization of artificially layered structures by epitaxial growth processes on semiconductors would seem to offer limitless possibilities for new materials.

Needs and Opportunities

Microelectronics is entering the submicron feature regime in which new classes of physical phenomena and materials structures determine device behavior. Materials demands can only be expected to increase. Materials development cannot be isolated from processing research and device design. Much greater interaction and open cooperation are needed among suppliers of materials, equipment vendors, and device manufacturers to work toward a common goal. As in other industries, integrated materials systems are needed.

Foreseeable changes in silicon technology will include flatter wafers, lower defect density (less than one defect per wafer), larger wafer diameter, and smaller feature size. To reliably achieve submicron circuit features, improved photolithographic processes will be needed. A combination of advanced optical methods (excimer lasers) and new materials (ultraviolet-activated photoresists and multilevel resists) will be used to produce features down to 0.3 to 0.5 μm. Below this size, soft x-ray and ion beam technologies will be needed. The United States is falling behind in these technologies, even though the phenomena on which they are based were developed in the United States.

In the growth of gallium arsenide (GaAs) and indium phosphide (InP), the primary needs are improved crystal purity, perfection, and size. An emerging technology that seems almost certain to become important is the epitaxial growth of GaAs on silicon and extensions involving optical detector regions (e.g., mercury-cadmium-telluride for infrared). A critical issue for the formation of a GaAs/InP industry will be the ability to transfer epitaxial growth processes from the laboratory to a manufacturing facility.

Research on II-VI semiconductors is less advanced than that on III-V materials. Important needs for improved materials are purer starting materials, crystal perfection, impurity doping techniques, uniformity of electrical and optical properties over large substrate areas, and effective passivation materials. The pursuit of heterojunction structures has recently begun and should offer significant opportunities.

Extensive numbers of process chemicals are used in the manufacture of semiconductor integrated circuits. They include solvents, acids, photoresists, gases for film deposition, film etchants and dopants, metal targets for conductors, and materials for device passivation. All of these materials have quality control problems. Among the problems are alkali and transition metal impurities, incorrect concentrations of intentional dopants, halogen impurities that cause corrosion, radioactive elements, and particles. Purity requirements will be even more stringent when submicron circuit feature size becomes standard.

Many improvements are needed in the polymeric and ceramic materials that are used in packaging electronic circuitry. Needed properties in polymers include a low dielectric constant, high thermal stability, high electrical resistance, and low moisture absorption. Similar properties are needed in ceramics, along with high thermal conductivity and thermal expansion coefficients similar to those for silicon. New ways of building integrated multilayer structures will require new and improved processes for deposition of metals, semiconductors, and ceramics. Layer-by-layer architecture will be supplanted.

Innovation in the laboratory and on the factory floor through processing research is the key materials opportunity and need in the electronics industry. As developers of new technology, we must extend the sophistication of our R&D laboratories into the factories where these new materials are made and converted into devices and systems.

Energy Industry

Scope of the Industry

The survey of the energy industry covered major energy production technologies now in use, as well as some developing technologies. The coal, petroleum, natural gas, and electricity industries are relatively mature, whereas the solar and geothermal production of energy is in an advanced phase of development. The nuclear power industry shares characteristics of mature and developmental industries in terms of needs for materials.

The use of energy from inanimate sources is central to every modern industrial economy. In the United States in 1987, the energy industry generated $376 billion in revenues and employed 1.23 million people. The 73.9 quadrillion Btu's of energy consumed in 1985 was distributed as shown in Table 2.10.

The price of energy is a significant factor in the performance of the U.S. economy, as shown by the economic downturns after the 1973 and 1979 oil crises. The importation of petroleum and petroleum products is a major adverse factor in the U.S. international trade deficit. In 1985 and 1986, the United States had a trade deficit of about $11 billion with the OPEC countries. This deficit is expected to grow as domestic oil production declines, energy demand rises, and petroleum prices increase.

Role of Materials in the Energy Industry

The energy industry is a conservative industry that relies mostly on conventional materials (e.g., steels, nickel-based alloys, and concrete). This trend should continue at least to the year 2000, with the traditional energy sources—coal, oil, gas, and nuclear—dominating the energy mix. Cost, service life, and reliability are dominant factors in the choice of materials for use in the various segments of the energy industry. The emphasis is on

TABLE 2.10   Distributions of Energy Consumed in 1985

Energy Source	Percent	Use Area	Percent
Petroleum	41.9	Electricity generation	35.0
Coal	23.7	Transportation	27.1
Natural gas	24.2	Industrial	23.8
Nuclear	5.6	Residential/commercial	13.3

evolutionary improvements in materials performance and reliability through improved refining, fabrication, and maintenance. New materials, however, can have a major impact on new energy systems (e.g., solar, advanced coal, and fusion) and can lead to improvements in established systems. Such improvements include titanium condensers, superalloy single-crystal turbine blades, and corrosion-resistant coatings in gas turbines.

Needs and Opportunities

The development of materials with certain economically desirable properties can influence the implementation of new energy systems and the efficiency of traditional systems. These need-generated opportunities are discussed below.

Transmission of electricity with conventional aluminum cable significantly limits the location of power plants. Development of practical superconducting materials that would function at liquid nitrogen temperatures would increase the efficiency of transmission over long distances and would permit much greater freedom in siting of new generating facilities. Practical superconducting cables could save about 75 billion kW of electricity worth at least $5 billion annually through elimination of transmission losses. Improved materials for generator cores and distribution transformers offer another major opportunity to increase efficiency and to reduce costs, perhaps by as much as $5 billion per year. Electric power plants (both fossil and nuclear fuel) will require life estimation and life extension to 60 to 70 years. Research is needed on methodologies for making such life estimates. Improved magnet and core materials for electric motors are another major area for improvements in efficiency worth billions of dollars. Two-thirds of electricity passes through some sort of motor drive, and 5 to 10 percent is lost in the process. Improvements in silicon processing technology would permit substitution of solid-state controls for mechanical speed control devices for motors. Savings of $4 billion to $5 billion annually might be achieved. In water-powered turbine-driven generators, there is an opportunity for life extension through improved materials for cavitation-resistant turbines. Improved insulating materials would have equally universal application.

In the area of coal combustion, the major opportunities are for corrosion- and erosion-resistant materials for combustion of coal in both conventional and advanced systems for generation of heat and electricity. If major coal-to-liquid fuel plants are to be implemented to conserve petroleum, many new demands for corrosion- and erosion-resistant materials must be met.

Resumption of commercialization of nuclear power will require solving many problems related to reliability of materials for reactors and to disposal of nuclear waste. Particular requirements are (1) a nondestructive measure of the condition of materials (surface and bulk) that is preferably on-line and continuous, and (2) a physically based mechanistic model of materials behavior in the environment. Implementation of fusion energy reactors (presumably well into the twenty-first century) will require solving a new range of materials problems.

Extracting petroleum and natural gas from ever harsher environments opens opportunities for new drilling equipment based on more durable and corrosion-resistant materials. Corrosion by carbon dioxide, hydrogen sulfide, and brine is extremely troublesome under the heat and pressure conditions of deep drilling. The depletion of easily accessible hydrocarbon deposits has resulted in a need for more sensitive detectors for prospecting.

Transmission and distribution of oil and gas require extremely reliable piping materials and joining techniques. New in situ inspection techniques could extend the life of existing distribution systems. The efficiency of petroleum transmission could be improved by flow enhancers. Improvements in the use of oil and gas can arise from improvements in catalysis. Significantly greater efficiency in the conversion of heavy petroleum to gasoline and jet or diesel fuel should be achievable through better catalytic cracking and refining processes. Catalysis for fuel cell operation can be a substantial factor in more efficient use of gas-based energy.

Wide use of geothermal energy, which is potentially accessible in much of the western United States and along the Gulf Coast, depends on reduction of capital costs for extraction. Improved materials performance could significantly increase the geothermal resource base and reduce geothermal costs, especially by reducing drilling and completion costs, improving plant and turbine durability, and improving plant on-line reliability. Many of the drilling and extraction materials issues are common to the oil, gas, and geothermal industries.

Converting solar energy into electricity is a problem with remarkably numerous solutions, including use of thermal-cycle generators, photovoltaic converters, wind-driven generators, ocean thermal gradients, and biomass conversion. All the technologies involved in solar energy conversion have materials limitations. The direct use of sunlight requires materials for collectors or concentrators that can perform their functions through 20 to 30 years of weathering. Photovoltaic conversion of sunlight to electricity on a commercial scale will require major increases in production of semiconductor materials such as monocrystalline silicon sheets, polycrystalline films, or amorphous films (Figure 2.5). Semiconductors other than silicon may have more appropriate electronic properties, but their use in devices will require major advances in processing the materials.

A number of desirable advances in materials processing and performance for the energy industry are identified in Table 2.11. These advances could have an impact across all components of the industry.

FIGURE 2.5 Solar concentration arrays employing GaAs photovoltaic solar cells. (Courtesy TRW, Inc. Copyright © 1986 by TRW, Inc.)

Metals Industry

Status of the Industry

The metals industry embraces the mining and smelting of metal-bearing ores and the refining and processing of metals into forms sold to other industries. The metals industry plays a central role in the U.S. economy; its products are found in almost every product sold in the United States. In the period from 1980 to 1985, the industry employed more than 1 million workers. Sales in 1987 totaled about $99 billion. These figures do not include the impact that the metals industry had on employment and sales in other industries (e.g., automotive and aircraft) that are large consumers of metals. The industry has a negative effect (e.g., −$10.8 billion in 1987) on the U.S. balance of payments.

The survey covered all the major metals or metal groups that were judged to be significant components of the U.S. metals industry. The metals and metal groups specifically represented in the survey were iron and steel, copper, lead, zinc, aluminum, refractory metals, titanium, and the nickel-based superalloys. The survey addressed both the traditional applications of metals and alloys and the newer applications such as metal matrix composites.

TABLE 2.11   Advances in Materials Processing and Performance for the Energy Industry

Major Advance	Impact of Major Advance	Critical Issues
Melting, refining, and/or deformation processing	Advanced alloys for use in corrosive/erosive environments and application Tough, embrittlement-resistant steels for elevated-temperature heavy-section application Single crystals for advanced turbines, solar photovoltaic converters, and power semiconductors Amorphous alloys for corrosion-resistant applications, soft magnetic properties, and solar photovoltaic converters	Optimum chemistry for specific fabrication process Development of stress-induced phase morphologies Purity and crystal perfection Fabrication of amorphous components
Powder consolidation	Transformation-toughened ceramics and high-purity ceramics for structural applications Near-net-shape forming for low-cost fabrication Mechanical alloying for high-temperature creep strength	Increase in temperature capability of transformation-toughened ceramics Production of submicron powders High cost and control
Advanced surface modification	Ceramic overlay coatings for enhanced environmental resistance Diffusion coatings for corrosion-resistant piping, tanks, etc. Coatings for improved wear and cutting characteristics Self-lubricating hard-facings for joints and valves Ceramic liners for high-temperature reactors and heat exchangers Artificial diamond and cubic boron nitride for drilling applications, wear-resistant surfaces, and semiconductor substrates	Coating integrity Adherence, porosity, durability, continuity Scale-up methodology Cost-effective processing Thickness trade-off vs. monolithic structure

Continued

TABLE 2.11   Continued

Major Advance	Impact of Major Advance	Critical Issues
Materials synthesis	Tailored complex inorganic materials for high-performance structural applications Unique nanostructures for electrical, magnetic, and superconducting applications Novel properties in nanoscale regime (supermodulus effect)	Molecular design of precursors Interactive modeling using supercomputers Scaling laws for microstructure-property relationships Development of processing methods for multilayers Availability of measurement and characterization tools
Life estimation and extension	Extend life of existing energy systems Improve reliability of existing systems Eliminate unexpected failures/forced outages	Validity of accelerated tests Realistic extrapolation methodologies Assessment of material condition online Retire/refurbish/replace decision methodology
Sensors and advanced characterization techniques	Monitoring of integrity of energy supply systems for improved reliability and performance Detection of underground energy resources (e.g., oil, gas, geothermal) System optimization and control by monitoring critical parameters In situ diagnostic capability, which reduces system downtime	Discrimination and validation of signals Expert decision making Sensors for hostile environments (extreme operating conditions) Signal transmission and recording Extend ultrasonic and x-ray tomography to full scale
Understanding erosion/corrosion mechanism(s)	Develop improved erosion-resistant materials and performance Modify system operations to avoid failures	Need for microscopic properties rather than global properties Establish relationships between local and global properties via micromechanical modeling

The metals industry has experienced severe economic difficulties over the last decade. The industry restructuring that has taken place in the last few years, combined with more favorable international trade conditions, has returned the industry to profitability, but at a generally lower production level. This lower production level reflects, in part, the import of metals and of metal-based products. For example, the intense import competition in the automotive industry has had a severe impact on the U.S. metals industry because car and truck production uses 15 to 18 percent of the steel, 12 to 16 percent of the aluminum, 8 to 10 percent of the copper, about 45 percent of the malleable iron, and about 60 percent of the lead consumed in the United States. The decline in metals production also reflects substitution of other materials such as plastics for metals in major products such as automobiles and consumer appliances.

The intense competition that resulted from an earlier overcapacity has reduced profitability and the ability of the industry to reinvest in both its physical and technical infrastructure. The survival of the U.S. lead and zinc industries is in doubt.

Role of Materials in the Metals Industry

The metals industry is a major supplier of materials to the U.S. economy. Metals are the prime construction material for the transportation industry and for large segments of the manufacturing, communications, energy, and construction industries as well as for military equipment. There is a strong interdependence between the robustness of these industries and that of the metals industry. The interdependence extends to research and engineering. The relationship is so strong that new materials may be developed by the user rather than the supplier of metals. In fact, in the lead and zinc sectors, new products must be developed by the user because the producers of the metals lack the research resources to do so. Some development is now done in other supplier industries such as the chemical industry.

New product development by metals producers includes a continuing search for materials with a high strength-to-weight ratio and for products that will have long lives in hostile environments such as seawater, brine wells, chemical process vessels, nuclear reactors, and the processing equipment of the metals industry itself.

The metals industry is also a major consumer of materials in the form of feedstocks such as ores and scrap metal, and the industry consumes huge quantities of fuel because the processing of metals is unusually energy intensive.

Needs, Opportunities, and Issues

In general, improvements in processing are essential to reduce costs and maintain a viable U.S. metals industry by achieving world-class cost-competitiveness. Critical processing needs of the industry derive from its role as a consumer of ores and other sources of metal. Technology for efficient processing of low-grade ores and for recycling of scrap metal and metal-containing wastes is clearly the most important processing need. To become cost-competitive in an international context, the steel, lead, and zinc industries need new processes to fully use their ore reserves in North America. The steel industry is actively pursuing the development of a strip casting technology to obviate the continuous casting and hot strip mill complexes. The lead and zinc industries need process technology specifically related to the characteristics of their ore deposits. A significant opportunity exists to fully exploit the value of silver and other minor constituents in minerals. The titanium and nickel industries lack the resources to implement innovative new technologies.

Secondary processing operations such as rolling and shaping need attention throughout the industry. Process simulation modeling, sensing, and control improvements are needed to upgrade quality and reduce costs. As an example of the potential benefits from sensing and control improvements, on-line inspection of hot steel slabs for surface defects could save $1.40 per ton. For 60 million tons of slab production, savings would be over $85 million. The payback period for the investment would be about 3 years.

Cooperative efforts of users and producers are necessary for effective product development. Product improvements will play an important role in maintaining market share in the competition of metals with polymers, ceramics, and composite materials.

There is an excellent opportunity for innovative research directed to accurately predicting metallurgical phenomena and product properties. This ability should be as quantitative as possible, hence the need for emphasis on models to relate process variables and product properties. Product properties studied should include the conventional mechanical properties in addition to those related to modern coating and joining processes. Fabrication technology, with emphasis on the applications of robotization and computer-integrated manufacturing, also needs significant attention.

An example of an innovative metal processing technique that is being applied to the formation of electrical transformer cores is shown in Figure 2.6. Rapid cooling causes the formation of an amorphous metal strip. Other amorphous metal (i.e., metallic glass) forms are shown in Figure 2.7.

Telecommunications Industry

Scope of the Industry

The telecommunications industry is a bellwether in the U.S. economy. It employed 1.01 million people in 1987 (904,000 in service and 103,000 in production of telephone and telegraph equipment). Sales of equipment amounted

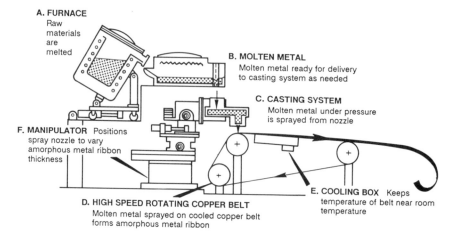

A. FURNACE
Raw materials are melted

B. MOLTEN METAL
Molten metal ready for delivery to casting system as needed

C. CASTING SYSTEM
Molten metal under pressure is sprayed from nozzle

F. MANIPULATOR Positions spray nozzle to vary amorphous metal ribbon thickness

D. HIGH SPEED ROTATING COPPER BELT
Molten metal sprayed on cooled copper belt forms amorphous metal ribbon

E. COOLING BOX Keeps temperature of belt near room temperature

FIGURE 2.6 Fabricating ribbons of amorphous alloy metal for electrical transformer core applications. (Reprinted, by permission, from *The New York Times*, Jan. 11, 1989, p. D7. Copyright © 1989 by the New York Times Company.)

FIGURE 2.7 Metallic glass product forms. (Reprinted, by permission, from Battelle. Copyright © 1989 by Battelle Memorial Institute.)

to about $16.1 billion; the service sector reported revenues of $130 billion. The downstream leverage of telecommunications is even greater—modern banking, airline travel, and commerce of almost every kind have been profoundly altered by the availability of inexpensive data manipulation and transport. Our modern information-based society results from a blending of the computer with telecommunications to provide the essential infrastructure for the information society.

The telecommunications industry is being reshaped by governmental deregulation in the United States and many other countries. International trade is increasing rapidly. Since approximately half of the world market is in the United States, foreign companies are trying vigorously to enter the U.S. market. The United States imports substantial amounts of communications equipment, which resulted in an unfavorable trade balance ranging from about $1 billion to $1.7 billion per year for the period from 1984 to 1987 (see Table 2.3).

Recent technology for data transfer and voice communication has blurred the distinction between the communications and the computer industries. The survey of the telecommunications industry emphasized the role of materials in equipment for telephone and data transfer operations. The findings parallel those of the electronic industry survey, except that a much greater role is played by optical technology in telecommunications.

Role of Materials in the Telecommunications Industry

The telecommunications industry is a major user and developer of new electronic and optical materials. In fact, it may be taken as a paradigm of a high-technology industry critically dependent on materials. As in the computer and electronics industries, most current devices are based on high-quality, dislocation-free, single-crystal silicon. The processing steps include masking, photolithography, diffusion, implantation, metallization, and etching. The packaging materials used to mount and seal the integrated circuits are commonly ceramics similar to those used in other electronic devices.

Quartz is crucially important to the industry for devices used for frequency control. Synthetic quartz has allowed the United States to become independent of overseas suppliers and is superior to natural quartz in cost, availability, and quality.

The shift from electronic to optical technology has required the development, production, and fabrication of many new materials. The development of new process technology resulted in silica optical fibers with transmission losses approaching the theoretical minimum. For optical emitters and detectors, III-V semiconductors are the materials of choice. Indium phosphide substrates with gallium-indium-arsenic-phosphorus epitaxial layers are used to generate radiation at 1.3 and 1.5 μm as input to optical fibers.

Needs and Opportunities

The telecommunications industry offers outstanding opportunities in materials science and engineering as the industry shifts toward optical technology. The United States has been a leader in this area, but vigorous effort will be required to maintain our position.

Achieving the economic opportunities inherent in the telecommunications industry is dependent on new materials. The modern central telephone office is essentially a large, special purpose, "switching" computer. A truly integrated optical switching system (all-photonic) with optical logic gates and memory could be much faster and cheaper than present electronic systems. It might be as great an advance as the electronic telephone office was over the electromechanical office.

The materials challenges to an all-photonic switching system are massive. Fully optical devices are scarcely at a laboratory prototype stage. Many promising photonic phenomena have been demonstrated, but current understanding and methods are inadequate to produce an all-optical system. Improved materials and processes are needed to make integrated optics a technological reality. Even in electrooptic systems, in which electrical fields are used to switch light signals, circuitry on III-V semiconductors has reached only a rudimentary level of integration because materials capable of a full range of functions are not available. For electrooptic switching, lithium niobate is useful, but better electrooptic materials that allow more elaborate switching circuits on smaller substrate areas would make this technology advance rapidly.

In the transmission of optical signals, silica-based optical fibers are approaching the theoretical limits of performance. If improvements in fiber transmission are to be made, new materials, possibly mixtures of metal fluorides, will be needed. In principle, fibers of these materials could provide transoceanic communications without repeater units because the optical losses are so low. Conversion to fluoride fibers will require new light sources and detectors, because transmission frequencies will move further into the infrared region.

A variety of new technologies will be needed if silicon very large scale integrated (VLSI) circuits are to be pushed to the limit of approximately 0.1 μm features. New families of photoresists, improved dry processing, new dielectrics, and new metallization technology will be needed.

Research Opportunities and Issues

Progress in the telecommunications industry has been a direct consequence of materials synthesis, processing, characterization, and analysis. Exceptionally promising areas for continued progress are superlattices, fibers, high-temperature superconductors, and neural networks (with circuitry analogous to that in the brain).

With techniques such as molecular beam epitaxy, precise control of semiconductor layers has been achieved. Despite this precision in fabrication, the chemistry and physics at the interfaces are scarcely understood. Multiple quantum-well structures have opened a new area of physics and have great potential for use in devices. The growth of III-V semiconductors on silicon and of II-VIs on III-Vs offers exciting new possibilities. Germanium-silicon superlattices may afford the opportunity to develop lasers and detectors based on silicon.

Optical fibers are now made at the theoretical loss limit. Devices such as Raman and Brillouin amplifiers and stress optic-effect sensors in which interaction lengths can be as long as kilometers should be possible. Single-crystal fibers of niobate, garnet, sapphire, and other oxides have potential, and electrooptic applications may follow.

Current research results on high-temperature superconductors are truly unprecedented. Although the fabrication of these oxide materials into useful configurations may be difficult, the promise of the materials is enormous, especially in tunneling technology and high-speed data busses.

In addition, biological information storage and retrieval systems hold great fascination. Addressing and reading out molecules are formidable tasks, but the scanning tunneling microscope may prove to be a useful probe.

SIGNIFICANCE OF MATERIALS SCIENCE AND ENGINEERING FOR THE PUBLIC SECTOR

Government—federal, state, and local—is critically dependent on materials in fulfilling its many missions related to defense, energy, transportation, space, and safety. While the materials science and engineering needs of the federal government are very diverse, they can be broadly divided into two regimes that often present quite different demands—one associated with providing new systems that can perform at the leading edge, and the other associated with incrementally improving the performance of existing systems.

Needs for leading edge systems exist in such programs as the strategic defense initiative and the national aerospace plane. In addition, more conventional planes need to fly faster and higher; submarines need to be faster and quieter and to have greater range; aircraft interiors should not burn following a crash; and computer capability is a frequent limitation in the ability to describe the behavior of complex systems and structures. In these and similar areas, successful development frequently depends on materials that have specific and definable characteristics either through inherent electrical, structural, or thermophysical properties or through engineering design that compensates for limitations of materials. While these demands are often difficult to achieve, the mission requirements present clear goals that the materials must meet.

Incrementally improving existing systems involves meeting numerous demands and setting goals that are often difficult to specify. Many federal units—including the Department of Defense (DOD), Department of Energy (DOE), Department of the Interior (DOI), Department of Transportation (DOT), Department of Health and Human Services (DHHS), National Aeronautics and Space Administration (NASA), and the regulatory agencies—are concerned with achieving optimum use and extended life of state-of-the-art structures, vehicles, processes, and devices. Many of the issues are common to several agencies. Corrosion limits the performance of ships, aircraft, bridges, concrete reinforcement, mining and drilling equipment, and vehicles of all types. The fuel efficiency of all vehicles can be improved by the use of lightweight materials. Energy availability and acceptability depend on developing reliable processes that are cheaper and less polluting than current processes. The useful lifetime of vehicles, tracks, roads, and structures can all be influenced by nondestructive testing methods.

In common with industry, the federal government needs improved generic techniques, such as synthesis and processing, and new techniques for the evaluation of performance. There are, however, significant differences in the impact of materials on the public sector and the private sector. Most importantly, national security is often a motivating factor for governmental involvement in materials science and engineering. The federal government devotes a large share of its materials R&D budget to the development of high-performance and high-cost materials for military applications that do not have a large impact on the civilian sector. The implications of this emphasis on defense-related needs are discussed in Chapter 7.

The following sections deal with materials requirements of four government units, illustrating needs for materials in the four areas of defense, energy, transportation, and space.

Department of Defense

Since the end of World War II, the United States has adopted the strategy that it will use technology to offset the numerical advantage that the Warsaw Pact nations have in manpower and conventional weapons. Materials science and engineering is intimately involved in maintaining the technological lead inherent in this policy.

The broad range of activities covered under the mission of DOD demands high performance from an exceptionally broad range of materials. Although hardware for weapons systems such as ships, tanks, and planes is most visible, materials for construction of buildings, roads, and runways; electronic and optical materials for communication and control; and clothing for climate and environmental protection are indispensable for many missions. Performance is, of course, paramount; but because the most severe demands occur during infrequent crisis situations, other considerations such as availability, reliability, and durability are important. Environmental protection is often necessary, and provisions for in-service inspection or evaluation are essential.

To provide more effective weapons and to minimize the risk to personnel, weapons systems are constantly provided with more intelligence. This intelligence requires microelectronics, sensors, displays, and software systems. Artificially structured materials and artificial intelligence are just two examples of areas in which research efforts support the diverse needs of the military.

In addition, present and future weapons systems are carefully examined to identify performance or economic issues that are limited by the capabilities of existing materials. When deficiencies are found, research programs are initiated to eliminate these limitations. The use of lightweight carbon-epoxy composite materials in military and civilian aircraft is an outgrowth of such an evaluation by the Air Force.

Many of the materials needed by DOD are unique to military applications. Therefore the development process must include the processes for producing materials, often to exacting tolerances. Strong attention is thus paid to manufacturing techniques.

Although DOD maintains numerous specialized laboratories and weapons centers, most of the actual R&D on new weapons systems is done by private contractors working under DOD direction. Many basic research programs in areas related to defense needs are carried out through contracts with universities.

Department of Energy

The Department of Energy has a mission that is much broader than its name suggests. It is responsible not only for R&D in support of advanced energy technologies and energy conservation, but also for the design, development, and production of nuclear weapons. It is therefore concerned with issues that are critical both to defense and to the economic well-being of the civilian sector.

Materials issues affect most DOE activities. DOE supports a large basic research program that provides scientific support for the energy technologies that are longer range than those described earlier for the energy industry. Examples of such long-range technologies are fusion power, advanced fission reactors, and improved techniques for coal conversion and combustion. Many of these new technologies are used in very severe environments that place unusual demands on materials.

The Department of Energy is also concerned with energy conservation issues, some of them near-term and others involving the development of longer-term technologies. These often involve materials and include establishment of standards for usage, development of new materials, and development of new systems that use both new and traditional materials. In the case of new energy sources for transportation, a major effort is under way to develop efficient electric vehicle power systems, including new and revolutionary batteries, new fuel cells, and ceramic turbines.

The Department of Energy is also charged with assuring that the United States continues to have the capability to maintain a credible nuclear deterrent, which includes the design and certification of new weapons and their production in stockpile quantities. For nuclear weapons to be a deterrent, they must threaten those targets that an enemy considers essential. Thus, as an enemy's strategy and its related targets change with time, so must the capabilities of our nuclear weapons, a condition that often requires development and qualification of new materials. For example, earth penetrator or nuclear-directed energy weapons present severe materials challenges. Hardening of weapons systems against enemy attack must be considered as well. Economic and environmental forces also have an impact on weapons research; for example, improved processing technologies for plutonium are required to reduce the amount of transuranic waste produced. The safe disposal of radioactive waste presents a series of unusually difficult and challenging materials problems.

The Department of Energy maintains several large national facilities for materials research, such as the National Synchrotron Light Source and several centers for neutron scattering research. Much of the short- and long-term research on materials is carried out in the multidisciplinary national laboratories supported by DOE.

Department of Transportation

The major needs of DOT are reflected in two mission requirements: enhancement of the safety of all modes of transportation and promotion of the efficiency of the transportation system. Improving the safety of the air traffic control system; ensuring the safety of bridges, pipelines, and ship hulls; establishing standards for performance of vehicles used in transportation; and providing for the safe transport of potentially dangerous substances all involve materials in a variety of ways.

Improving the efficiency of the transportation system also involves materials and materials systems. The efficiency of the vehicles that move along the nation's highways, airways, waterways, and railways is continually being improved as a result of the introduction of new materials into their power plants and structures. The efficient operation of the transportation system also depends on the ease with which freight and humans are transferred to vehicles. Efficient transfer depends significantly on the creation of efficient interfaces between the various modes of the transportation system. Since these interfaces are frequently concerned with handling materials, they also are critically dependent on the progress that results from research on materials. It is also worth noting that the need for control of large amounts of information, often in real time, places increasing demands on the control systems and the computers that support these systems.

From the materials science and engineering perspective, the main materials of interest can be classified according to type and weight. They are (1) bulk materials such as concrete, asphalt, and aggregate, which form the overwhelming percentage of the weight of railbeds and roadbeds; (2) structural metals (primarily steel), which perform critical structural functions in the form of rail tracks, railcars, bridges, ships, pipelines, and concrete reinforcement and in new metal alloys and reinforced plastics that are being used increasingly in aircraft and ground vehicles; and (3) specialized materials such as polymeric materials used for protective clothing, coatings, adhesives, and vehicle interiors and exteriors.

Although the relevant missions of DOT have been categorized as safety and efficiency, these two functions are clearly intertwined. Deterioration, when unchecked, leads either to unsafe conditions or to less than optimal use of infrastructure capacity. Overall evaluation of the needs and opportunities for research in the U.S. transportation sector indicates that new materials technologies offer substantive possibilities for improving all areas related to the transportation infrastructure.

National Aeronautics and Space Administration

The National Aeronautics and Space Administration is charged with assuring that the United States remains a preeminent world power in space science, space operation, and space exploration, and with creating the necessary research and technology bases needed for the United States to maintain a competitive position in civilian and military aeronautics.

Materials issues pervade all aspects of NASA's mission. Many of the challenges are unique and particularly difficult. Special emphasis is placed on weight reduction to enhance payload capabilities. This includes not only lightweight materials, but also innovative processing and design concepts that can reduce weight while maintaining the desired level of performance. High-temperature materials are required to enhance fuel efficiency in combustion and to ensure protection during reentry. Cryogenic materials and insulation are needed for containment of cryogenic fuel and liquid oxygen. High-efficiency energy sources for space applications present critical materials problems. Environmental conditions include the high-vacuum conditions of outer space and the effects of atomic oxygen encountered in low earth orbit. Reliability is an important issue, since maintenance is often difficult or impossible. NASA is also responsible for exploring effects of the space environment, including microgravity, on materials processing.

The National Aeronautics and Space Administration also has a special role in assuring the continuing availability of advanced technology for ground-based aeronautics systems. This includes concern with improving the efficiency of turbine engines, the design of low-noise, high-thrust turbines, the examination of new structural designs that increase lift and reduce weight, and the development of new avionic systems that improve reliability and provide needed assistance to their human operators.

Materials issues thus are of critical concern to the mission of NASA. Economics, safety, and performance all depend on innovative use and development of materials. To meet the myriad demands for advanced sensors for earth-based and planetary missions; for on-orbit and deep space power; for micro- to mega-thrust propulsion systems; for extraterrestrial structures and vehicles; and for improved airframes and propulsion systems for the commercial aircraft industry, NASA supports a wide spectrum of materials research in its in-house laboratories and through contracts with industry. NASA also maintains a wide range of unique test facilities that are used by government and industry for the evaluation and testing of new designs and concepts.

FINDINGS

The surveys of the eight industries show critical needs of these industries for new, improved, and more economical materials and processes. Similar needs are evident in the public sector in areas including defense, energy, transportation, space, and health. Some important crosscutting materials needs are summarized in Table 2.4.

An overriding theme for all the industries surveyed was the primary importance of synthesis and processing of new materials and traditional materials, and fabrication of these materials into useful components and devices. Materials science and engineering, and processing in particular, plays a uniquely important role in these industries and in their ability to help maintain and improve the U.S. position in international competitiveness.

In every industry surveyed, there is a clear need to produce and fabricate new and traditional materials more economically, and with higher reproducibility and quality, than is done at present. The opportunities in synthesis range from preparation of totally new materials to development of methods for recycling scrap metals and polymers. For all materials, there is an acute need for better ways to produce objects in shapes approaching the desired final form (near-net-shape forming). Equally important is a need for ways to determine the quality of products on the production line and to feed back

PLATE 1

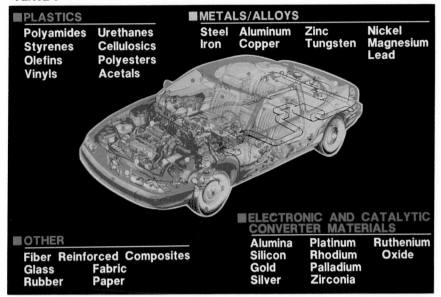

PLATE 2

PLATE 1 View of a modern automobile showing the location of various types of materials in the structure of the body and components. (Reprinted, by permission, from General Motors Research Laboratories, 1989.)

PLATE 2 Plasma immersion implantation chamber for surface modification. (Courtesy TRW, Inc. Copyright © 1986 by TRW, Inc.)

PLATE 3

PLATE 3 Centrifugal atomization, one of several techniques for creating samples of rapidly solidified metal. (Reprinted, by permission, from Bernard H. Kear, 1986, Advanced Metals, Sci. Am. 255:159–167. Copyright © 1986 by Pratt & Whitney Aircraft.)

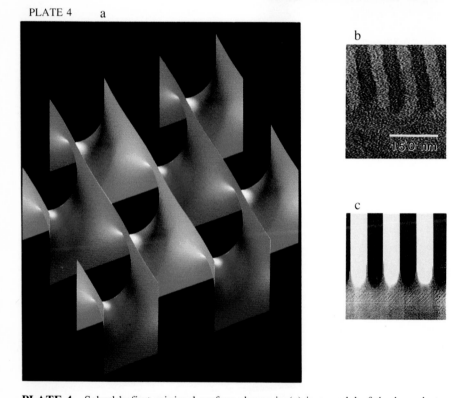

PLATE 4 a

b

c

PLATE 4 Scherk's first minimal surface shown in (a) is a model of the boundary between two polymers in a copolymer. An electron microscope image of an actual copolymer (polystyrene-polybutadiene) shown in (b) matches a projection (c) of Scherk's surface. (Reprinted, by permission, from Edwin L. Thomas, David M. Anderson, Chris S. Henkee, and David Hoffman, 1988, Periodic Area-Minimizing Surfaces in Block Copolymers, Nature 334:598–601. Copyright © 1988 by Macmillan Magazines Ltd.)

the information to operators in real time. Computers and computer modeling are beginning to play an important role in this area and are also reducing the time needed to take new materials, processes, and designs from the laboratory to the production floor. Throughout manufacturing, the integration of synthesis, processing, fabrication, and testing is a challenge to materials science and engineering.

Industry needs in other areas of materials science and engineering are also evident from the industry surveys outlined above. These include needs for new materials with new properties, especially composites, and new materials synthesized at the nanostructural level. Outstanding opportunities exist to improve the production and use of materials through computation as a complement to experimentation. A challenge is to use our understanding of structure, bonding, and properties to develop predictive models from the behavior of materials in use. There is a critical need for better ways to predict the mechanical behavior and useful lifetime of materials and objects in various applications.

Another theme that emerged in some of the industry surveys was that the federal government should help to identify industries of current or projected strategic national importance. Such identification should then influence the emphasis and direction of major national research activities. It was also felt that the government should play a role in helping to bring industry, universities, and federal laboratories together to address these research priorities.

The industry survey participants saw a number of opportunities to improve the effectiveness of the various institutions involved in materials science and engineering. Their views, and those of the committee as a whole, represent important themes of this report and are as follows:

• *Industry* clearly has the major responsibility for maintaining the competitiveness of its products and its production operations. Greater emphasis on materials science and engineering and, in particular, on integration of materials science and engineering with other business operations is necessary to improve the competitive positions of U.S. firms in domestic and international competition. The incentives (e.g., money and prestige) for top-quality people to become involved in production should be increased. Intelligent collaborations with researchers in the universities and in government laboratories can enhance the effectiveness of R&D in industry. Industrial consortia can provide a mechanism to conduct R&D programs too large for any one company.

• *Universities* traditionally have had a dual role in educating personnel for industry and in conducting innovative fundamental research. The universities can promote the general welfare through encouragement of the interdisciplinary teaching and interdisciplinary research characteristic of materials science and engineering. Both industry and the nation need materials scientists and engineers broadly trained in the range of disciplines needed for effective research, development, and production. Greater emphasis is needed on teaching and research relevant to processing and manufacturing operations. Universities will often provide the best sites for the science and technology centers discussed below.

• *Government laboratories*, which include hundreds of laboratories funded by federal and sometimes by state governments, have many capable employees and large capital resources that could benefit industry. The DOE-funded national laboratories, in particular, have many scientists and engineers with special talents in materials science and engineering. Reorientation of the missions of the national laboratories toward industrial materials science and engineering interests could have a valuable effect on U.S. industrial competitiveness. (The role of the National Institutes of Health laboratories as an asset to the pharmaceutical industry is illustrative.) To be effective in helping industry, federal R&D must be directed intelligently to problems of genuine interest to industry. Exchange of personnel between industry and the government laboratories would help focus the work and assist technology transfer. The federal laboratories, especially the National Institute for Standards and Technology in its new role, could play a valuable role in establishing test procedures, setting standards, assembling data collections, and transferring technology to industry.

• *Centers*, including the materials research laboratories and engineering research centers funded by the National Science Foundation, play an important part in bringing together people from the many science and engineering disciplines that constitute materials science and engineering. Likewise, centers can be a focal point for bringing together people from universities, industry, and government laboratories in materials science and engineering programs of mutual interest. This combination of many talents and the extensive instrumentation and equipment available in a center can be extremely effective in advancing R&D programs. Relevant models exist in the interdisciplinary teams at large industrial laboratories and at the Max Planck and Fraunhofer Institutes in West Germany. As with the national laboratories, significant input from industry will be needed to direct centers in work relevant to industry.

• *Consortia* of industrial research groups, such as the Microelectronics and Computer Technology Corporation and the Semiconductor Research Corporation, and consortia in steel, machine tools, and components may play a significant role in preliminary research on new technologies such as "beyond VLSI" circuitry. There are few U.S. models of demonstrated effectiveness. Industrial consortia guided by the Ministry of International Trade and Industry seem to be effective in Japan but need refinement as models for U.S. use. Leadership and initiative to establish such mutual materials science and engineering ventures should be encouraged.

- *People* who are well trained and well motivated, and who have effective leadership skills, are the basis for success in any technological endeavor. The universities, industry, and government all have important roles in ensuring the availability and intelligent employment of materials science and engineering personnel. Communication and personnel interchange are fundamental to successful technology transfer within a company or between institutions. All parties in materials science and engineering should work to encourage communication and a sense of community in ventures aimed at enhancement of U.S. industrial competitiveness.

3

Research Opportunities and Functional Roles of Materials

This chapter describes research opportunities in materials science and engineering in terms of the functional roles of various classes of materials. Although such an approach emphasizes the connection between basic research and applications, the discussions in the following sections focus on the scientific and technological frontiers that lead to development of new materials and to processing of new and conventional materials. The power of materials science and engineering is most apparent when it combines intellectual, scientific, and technological opportunities at these frontiers, while also addressing societal needs.

The practical exploration of research on new materials and processes has grown increasingly effective as understanding of composition, microstructure, and properties has progressed, as the degree of control over these characteristics has increased, and as theoretical models have been applied to increasingly complex problems. Materials researchers can now analyze and manipulate the properties of materials in ways that were hard to imagine just a few years ago.

At the atomic level, instruments such as the scanning tunneling microscope and the atomic resolution transmission electron microscope can reveal, with atom-by-atom resolution, the structures of materials. Ion beam, molecular beam, and other types of equipment can build structures atom layer by atom layer. Instruments can monitor processes in materials on time scales so short that the various stages in atomic rearrangements and chemical reactions can be distinguished. Computers are becoming powerful enough to allow predictions of structures and of time-dependent processes, starting with nothing more than the atomic numbers of the constituents.

At higher levels of structure, researchers are beginning to understand, and build, structures with crystals or "grains" that contain only small groupings of atoms, in which as many atoms lie in the grain boundaries as in the grains themselves. Researchers are also finding new properties in "nanocomposites"—composites on the scale of nanometers. Element sizes in electronic chips are rapidly decreasing and are approaching the size of small groups of atoms.

The level of microstructure and macrostructure above the nanometer scale continues to have rich promise for research. Developments at this level include modern composites, directionally solidified high-temperature turbine blades, and flaw-tolerant ceramics. Development and application of modern fracture mechanics and design have also been important. Much of the innovation and development in modern processing is concerned with controlling structure at this level as well as at finer levels. Examples include new strip casting processes and near-net-shape forming. Major opportunities exist for computer modeling to aid development of new processes and more rapid introduction of new designs and novel production processes.

The following sections describe selected research opportunities for structural, electronic, magnetic, photonic, and superconducting materials and for biomaterials. The list is by no means intended to be complete, but rather to illustrate the vitality and rich promise of the field. The emphasis on function underscores the use of materials; it makes explicit the link between fundamental research and the applications of research; it highlights the opportunities for research to contribute to areas of societal need.

STRUCTURAL MATERIALS

The properties of structural materials—toughness, strength, hardness, stiffness, and weight, for example—are determined by the interaction of atoms by their arrangement as manifested in molecules, crystalline and noncrystalline arrays, and defects, and by higher levels of structure including flaws and other microscopic heterogeneities. Consequently, the ability to predict and control materials structure at all levels, from the lattice dimensions to the macroscopic level, is central to developing structural materials that achieve the level of performance needed to accomplish the nation's technological, economic, and military goals.

Two examples bring this relationship into focus. Without major advances in propulsion technology, reducing the structural weight of launch vehicles is the only means of lowering the prohibitive cost of transporting heavy commercial, scientific, and military payloads into orbit. Reducing structural weight will require new materials with very high strength-to-weight ratios and the capacity to withstand high temperatures. Currently, the materials used to make the structural components of the Space Shuttle and today's expendable rockets are essentially improved versions of those used in the Apollo program. These materials are approaching their technical limits. Given that the gap between the discovery of a new high-performance material and its actual production may be as great as 15 years, today's basic research should be acquiring the knowledge needed to design and fabricate the structural materials required for aerospace applications during the first decades of the next century.

Similarly, satisfactory disposal of high-level radioactive waste will depend, in large part, on R&D in materials science and engineering. Plans now call for incorporating waste materials into a relatively insoluble material, probably a glass or a ceramic. Then the integrated mass will be encased in a canister that, first, must withstand any potential accident during shipment to a waste repository and, second, must remain impervious for centuries. The canister and the waste form itself must be able to withstand the assaults of corrosion from the outside and the attacks of radiation from the inside. The search to identify and fabricate materials that can satisfy these requirements is under way, but fundamental understanding of the properties of materials and of the forces that degrade them and reduce their performance would advance these efforts.

Over recent decades, the development of structural materials has evolved from an activity guided almost entirely by empiricism to one in which theory is playing an increasingly important role. The design of alloys, for example, has benefited greatly from the application of fundamental principles. Yet theoretical inputs to these efforts and those addressed to other materials classes are largely in the form of qualitative guidelines. Given the complexity of most materials systems and the complexity of the mechanisms of deformation, fracture, and degradation, this is not surprising. The opportunity now exists, however, to fill important gaps in understanding and to develop theories that provide quantitative guidelines for the design of materials. With new instruments such as the scanning tunneling microscope and the atomic resolution transmission electron microscope, it is possible to view defects on an atomic scale. Supercomputers can now perform the many calculations required to determine the properties of a particular combination and arrangement of atoms. Further, process modeling and incorporation of novel processing techniques permit creation of some microstructures under reproducible conditions. Using these tools, the materials scientist is busily designing experiments that address fundamental questions about materials systems, the answers to which can provide the information necessary for erecting a unifying theoretical framework that fosters discovery.

The examples below highlight some important issues arising from basic research on structural materials, arranged according to materials class. The spectrum of structural materials includes metals, ceramics, polymers, and composites. In a complex system, it is common to have niche applications for each of these.

Metals

The central role of processing in controlling and improving the performance of structural materials, first recognized in the late 1960s, has become a major focus in this area in the past decade. Processing is now viewed as complementary to the chemistry and physics of materials as a tool in controlling properties.

The advent of new materials, computer controls, and new sensing devices in combination with the use of process modeling and the concepts of artificial intelligence, as well as other new technologies, has permitted the recent development of a wide range of important and exciting new processing methods. These new processes have had an enormous impact on the cost and quality of materials. Rapid and efficient adoption of such processes has significantly affected the economic well-being of industries and countries that have adopted them, and continuing important developments in this area can be expected.

In the case of metals, continuous casting is an important example of an exciting new processing method developed in the past two decades, and extension of that technology to thin-slab and strip casting is the focus of much current effort. The broad area of application of magnetohydrodynamics to metal (and semiconductor) production and fabrication is still a largely unexplored field. Recent and past successes in the area include electromagnetic stirring, dampening of convection, moldless continuous casting, levitation, pumping, containment, and forming. New, important net shape forming processes for metals include injection molding, semisolid forming, innovative and economically promising investment casting techniques, hot chamber die casting, and spray forming. The last technique can combine the benefits of net shape processing and rapid solidification processing.

In the field of surface treatments and coatings, newer plasma processes permit large-area deposition of a wide range of coatings for decorative, protective, or functional purposes. Laser surface treatment (Figure 3.1) and laser cutting and shaping are being used more widely.

Important research opportunities exist in metal processing, and processing improvements are essential to reduce costs and maintain a viable U.S. metals industry. Critical processing needs of the industry derive from its role as a consumer of ores and other metal sources. Technology for efficiently producing high-quality metals and semifinished shapes from primary ores and from recycled scrap metal and metal-containing waste materials is clearly the most important processing need. To become cost-competitive in an

FIGURE 3.1   Laser and material interaction during welding. High-power carbon dioxide lasers provide a unique thermal energy source for processing materials. This energy source is precisely controllable in intensity and position. Laser beam welding is currently being used to fabricate components for naval structures and platforms, and its use is expanding rapidly. (Courtesy Naval Research Laboratory.)

international market, the steel, lead, and zinc industries need new processes to utilize fully their ore reserves in North America. The lead and zinc industries also need processing technology specifically related to the characteristics of their ore deposits. A significant opportunity exists to fully exploit the value of silver and other minor constituents in minerals.

Secondary processing operations such as rolling and shaping need attention throughout the metals industry. Process simulation modeling, sensing, and control improvements are needed to upgrade quality and reduce costs. New opportunities now exist in these areas as a result of developments in related technological fields.

Cooperative activities involving both users and producers are the most effective way of developing metal products. Product improvements will play an important role in maintaining market share in the competition of metals with ceramics, polymers, and composite materials. There is an excellent opportunity for innovative product research that will enable accurate prediction of metallurgical phenomena and product properties. This ability should be as quantitative as possible, hence the need for emphasis on models to relate process variables and product properties. Product properties studied should include the conventional mechanical properties in addition to those related to modern coating and joining processes. Fabrication technology, with emphasis on the applications of robotization and computer-integrated manufacturing, also needs significant attention.

Researchers working on new metallic materials can now also focus on innovative processing techniques to produce "materials by design." In this approach, the alloy designer can anticipate being able to engineer metals alone or in combination with ceramics, or in different chemical states, to optimize performance. Examples include the following:

- *Very fine grained, single-phase materials.* These are achievable through advanced processing techniques, such as rapid solidification processing, permitting the attainment of structural refinement and control of metastable and equilibrium-phase transformation. Such materials can have excellent combinations of strength, ductility, and corrosion resistance.

- *Dispersion-strengthened and microduplex alloys.* Again, advanced techniques such as mechanical alloying can create unique microstructural configurations leading, for example, to ideal multiaxial properties and microstructures that are stable at high temperatures.

- *Intermetallic compounds.* Unique combinations of heretofore difficult-to-form compounds are now possible in polycrystalline, single-crystal, and amorphous forms by the use of novel processing techniques as well as compositional control and trace element additions. These promise to allow for higher-temperature, load-bearing applications.

- *Composites.* Innovative combinations of metal matrix, intermetallic matrix, and ceramic matrix composites are clear examples of materials produced by design. Again, processing is the controlling parameter.

- *Thin films and layered structures.* These are, perhaps, the ultimate examples of creating unique microstructures, literally by atom-by-atom layering.

- *Alloys resistant to radiation damage.* Surface modification is one technique for developing such alloys. Other approaches based on ceramic or polymer matrices cannot be implemented until researchers learn how to design systems with less ductile materials and to incorporate them into structures. To accomplish this will require a multidisciplinary approach, including calculation (theory-assisted engineering) and modeling, particularly of micromechanical aspects.

Achieving the ideal microstructure for a particular materials system requires fundamental understanding of the many mechanisms of failure—brittle and ductile fracture, fatigue, creep, friction, and wear. Each is the product of complex interactions, but mathematical models of failure mechanisms are limited to simple two-dimensional systems or are based on averages that mask many important details. These models must be extended, and new ones developed, to account for the interaction of atoms in three dimensions. The use of the whole arsenal of characterization tools and a complete understanding of synthesis-processing-structure-property relationships will also be needed. It is anticipated that the latter, in particular, will be aided by artificial intelligence and expert systems and by the development of reliable data bases.

Ceramics

Not so many years ago, ceramics processing comprised predominantly separation and crushing of naturally occurring minerals, followed by sintering. Today, many advanced and even conventional ceramics are produced chemically from pure materials with vapor processing, aqueous precipitation, or sol-gel techniques. Melt processing, melt alloying, and melt refinement—followed sometimes by rapid solidification—are used in producing a wide range of ceramic materials, including abrasives. Hot isostatic pressing, forging, and extrusion of ceramics are now, or will be, playing an important role in forming ceramics.

New processes developed specifically for advanced structure control include many of the above processes. Ultrafine grain sizes, useful in a wide range of materials, are obtained (depending on the part or material) by condensation from the vapor, controlled rolling, heterogeneous nucleation, and electromagnetic stirring. Amorphous or glassy structures are obtained by rapid solidification processing, vapor deposition, electrodeposition, and other solidification processes that operate far from equilibrium. Oriented grains or crystals are obtained by a wide range of processes to control crystallization or recrystallization behavior. Processes have also been developed to control structure at the level of the lattice spacing. Low-dislocation or dislocation-free single crystals are now commonly grown from the melt or are obtained through subsequent processing; much improvement in the quality of such crystals is needed and can be expected in the years ahead.

The appeal of ceramics as structural materials is easy to understand. Ceramics are light, but their compressive strength matches or exceeds that of metals; they can withstand extremely high temperatures; they are exceptionally hard, are resistant to abrasion, and are chemically inert; and they excel as electrical and thermal insulators. If research can add two other properties to this list—tensile fracture toughness and ease of processing—ceramics are likely to become ubiquitous in structural applications.

How rapidly ceramics will achieve their potential will be determined largely by research on processing techniques, which first must overcome the problem of brittleness and then must prove to be economical. One processing requirement is either to eliminate defects—the voids, agglomerations, and chemical impurities from which cracks originate—or to toughen ceramics by devising ways to prevent cracks from spreading.

Several approaches yield pure, minute particles, the starting materials needed to achieve a void-free microstructure during synthesis. Attrition, melt atomization, and physical vapor deposition produce such particles without chemical change. Chemical precursors are involved in sol-gel and thermal degradation processes.

Dense ceramic bodies are usually obtained by pressing and sintering of fine particles. Modern processes include hot isostatic pressing and forging, which apply pressure during the sintering process. Dense bodies are also formed by direct casting of the melt and by processes that involve oxidation of a melt to obtain a ceramic or metal-ceramic composite. Infiltration of a ceramic precursor with a melt or vapor is another method used, and vapor forming is yet another. In self-propagating, high-temperature synthesis, reactive species are heated in a mold under pressure and are allowed to heat to a temperature at which they react and densify.

Developing tough, strong ceramics is another important, major goal of much ceramics research today. Transformation toughening is one important method, a practical example of which is seen in ceramics whose structure is composed of at least partly tetragonal zirconia. As a crack begins to grow in such ceramics, the tetragonal structure becomes monoclinic, with a resulting volume change of 3 to 5 percent that arrests crack growth. Fifteenfold increases in toughness have been achieved in such materials, and important structural ceramics and abrasive materials are built on these compositions. Other ways of improving the toughness of ceramics involve achieving and maintaining very small grain sizes within the ceramic body, incorporating controlled and dispersed voids, and synthesizing ceramic-ceramic composites. Toughening mechanisms in a ceramic are illustrated in Figure 3.2. An example of a ceramic composite (silicon carbide in alumina) is shown in Figure 3.3.

Research on new toughening mechanisms, as well as on new processing technologies, offers much promise for the future. Concomitant advances in nondestructive testing will also be needed. X-ray tomography, ultrasound, and other existing nondestructive testing methods must be refined to improve flaw detection, and new methods should be developed for ceramics systems. Probabilistic design methodology can play an important role in predicting the performance of ceramic parts. Modeling efforts, however, will require statistically valid data on crack propagation and the stress behavior of ceramics.

One exciting new development is the growth of diamond or diamondlike materials on the surface of various substrates. Whereas previously diamonds were grown at high pressures and temperatures, it is now possible to grow

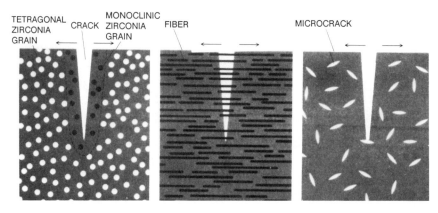

FIGURE 3.2   Toughening mechanisms, by which a crack in a ceramic can be arrested, complement new processing techniques that seek to eliminate crack-initiating imperfections. Transformation toughening (left) relies on a change in crystal structure that zirconia grains undergo when they are subjected to stresses at a crack tip. Ceramics can also be made crack-resistant by interlacing with fine ceramic fibers (middle), as is the case in composite materials. A third way to stop a crack is by spreading the stresses concentrated at its tip over a larger surface. This can be accomplished if minute cracks, called microcracks (right), are purposely created in the ceramic material during processing. (Reprinted, by permission, from H. Kent Bown, 1986, Advanced Ceramics, Sci. Am. 255:169–176. Copyright © 1986 by Scientific American, Inc. All rights reserved.)

crystalline diamond films by energetically assisted chemical vapor deposition processes at far lower temperatures and at low pressures. Diamond films may find application as wear-resistant coatings and as bearing surfaces. Their high thermal conductivity already has led to use of diamond materials as heat sinks in electronic applications, while their combined thermal conductivity, low expansion coefficient, and strength provide high resistance to thermal shock, a characteristic desirable for instrument windows. Applications for diamond films will expand rapidly in the future as the availability of large-area films increases.

Polymers

Polymers are notable for the unique combinations of properties that are possible. To the mechanical attributes of high strength, high flexibility, and light weight can be added thermal, electrical, and optical properties that make polymers especially well suited for specialized applications. In defense applications, for example, tough, lightweight polymers are likely to be used as armor, protecting against projectiles and lasers while offering the additional benefit of radar transparency. In the last decade, polymers have emerged from their early role as primarily inexpensive commodity materials into the value-added, high-technology sector in which specific properties can be exploited for novel performance. Examples include piezoelectric polymers used for ultrasensitive sonar, high-temperature flame-retardant fabrics, nonstick surfaces, reusable pressure-sensitive adhesives, and encapsulants for drug delivery.

Structural polymers are already in widespread use and account for a major portion of the $100 billion global market for polymers. However, improvements in existing materials and development of new polymers with significantly enhanced properties are possible and will increase the range of structural applications and the size of the market. Progress to date is well illustrated by the remarkable strength-to-density ratio of aromatic polyamide polymers; development of other impressively strong polymers of lower cost can be expected. Polymer adhesives are already replacing rivets in many joining applications, and polymer-fiber composites are replacing metals in many

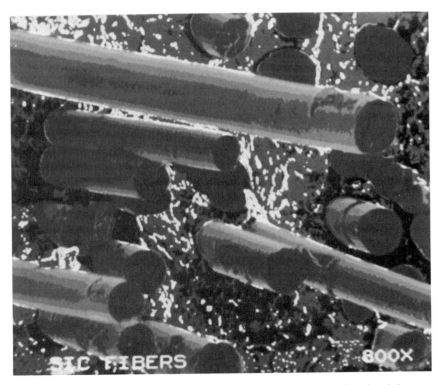

FIGURE 3.3 Fracture surface of silicon carbide-reinforced alumina. (Reprinted, by permission, from E. I. du Pont de Nemours & Co., Inc. Copyright © 1989 by E. I. du Pont de Nemours & Co., Inc.)

structural applications. Finally, polymer and polymer-fiber composite materials are finding extensive biomedical applications. All of these trends are expected to continue, and all present research challenges in synthesis and processing to achieve desired structures, properties, and performance.

Advances in computing power offer opportunities for substantial and badly needed improvements in the theoretical underpinning of polymer science. These include the modeling of stable and metastable structures of macromolecules, of liquid crystallinity, of phase relations and phase-transition kinetics (including the now poorly understood glass transition in amorphous materials), and of nonlinear viscoelastic behavior in polymer melts (which is critical in polymer processing). The recently developed reptation concept is an interesting approach to the last problem; identification of the collapse transition of polymer solutions by use of computer-intensive Monte Carlo methods, combined with renormalization and scaling theories, illustrates the benefit of increasing computing power in modern theory.

The long-chain connectivity inherent in polymers imposes unusual constraints at or near interfaces, where abrupt changes in density, composition, and orientation can occur. Understanding polymeric interfacial problems requires sophisticated scientific studies; scaling up from analyses of small molecules will not suffice, because the behavior of polymer chains can be markedly different from that of low-molecular-weight compounds. Studies should address the physical and chemical dynamics of surfaces, interfacial regions of polymer blends and alloys, polymer melts in confined geometries, and crystal-amorphous interfaces in semicrystalline polymers. These interfaces control adhesion, wetting, colloidal stabilization, and mechanical properties, and they are intimately involved in the failure and deformation of polymers. Yet experimental and theoretical investigation of polymer interfaces has begun only recently.

Chemistry is important to all the materials classes, but none benefits more from close interaction with the field of chemistry than does polymers. The synthetic capabilities of the modern chemist can be advantageously applied in the rational design of molecules to achieve desired properties. Specificity in molecular composition, architecture, and size can be tailored with amazing precision. The attachment of functional groups at selected locations is readily accomplished. Moreover, control of chain length and composition provides for extremely well-defined materials that can self-assemble into intricate patterns sometimes mimicking those of biological systems.

Interfaces are also of crucial importance to the performance of composites (see below), many of which use polymers for the matrix, the reinforcing dispersed phase, or both. Defects or weaknesses at interfaces can severely limit the performance of composites. Figure 3.4 shows a carbon fiber-reinforced composite fracture surface. Recent research on polymers, however, suggests a route to eliminating interfaces while achieving many of the su-

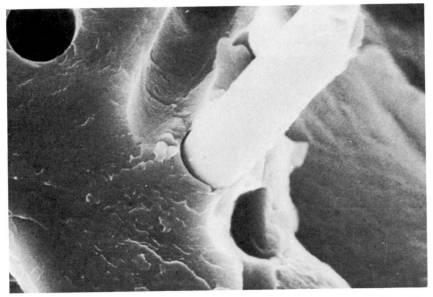

FIGURE 3.4 Micrograph of carbon fiber-reinforced composite fracture surface. (Reprinted, by permission, from E. I. du Pont de Nemours & Co., Inc. Copyright © 1989 by E. I. du Pont de Nemours & Co., Inc.)

perlative properties that composites can offer. Now under investigation are methods to achieve molecular mixing in which rigid backbone molecules are uniformly dispersed in a matrix of flexible-coil polymers. The result is a "molecular composite," akin to a fiber-reinforced composite with the individual rigid macromolecules acting as the "fiber." The uniform dispersion creates a polymer mixture of long-range molecular order and, thus, with no interfaces. Molecular composites are expected to have superior impact resistance, fracture toughness, and compressive strength, while offering opportunities for novel combinations of electrical and optical properties.

Research on molecular composites is in the very early stages. Progress will depend on finding answers to questions about the nature of liquid-crystalline polymers. The physics of the isotropic-anisotropic phase transition must be elucidated, and much more must be learned about chain conformations and intermolecular packing in mesomorphic states and about the effects of chemical structure and chemical reactions between chains on long-range order. Molecular composites are but one of many exciting technologies that are likely to emerge from fundamental research on polymers.

Composites

As should be apparent by now, the traditional categories for classifying materials can impose artificial distinctions. Although some research issues are unique to one class of materials such as metals, many span the categories of materials or stem from technological applications that exploit the properties of a range of materials. Nowhere is this overlap of scientific interest and resultant benefits more apparent than in the area of composites.

Loosely defined, composites are hybrid creations made of two or more materials that maintain their identities when combined. The materials are chosen so that the properties of one constituent enhance the deficient properties of the other. Usually, a given property of a composite lies between the values for each constituent, but not always. Sometimes the property of a composite is dramatically better than that of either of the constituents. The potential for such synergy is one reason for the tremendous interest in composites for high-performance applications.

Understanding and control of the chemical changes that occur during creation of composites will be critical to achieving the enormous potential utility of these materials. This is a formidable challenge, requiring the contributions of experimentalists and theoreticians in metallurgy, polymer science, ceramics, organic and physical chemistry, rheology, and solid-state physics. Determining the molecular changes that occur in bulk materials during processing, which involves many steps, will be an extraordinarily difficult problem. The chemical dynamics of curing polymeric matrix composites is fairly well understood, at least in a qualitative sense, but very little is known about the reactions occurring during the production of metal matrix and ceramic matrix composites. Filling these serious gaps in understanding would advance efforts to develop fabrication methods to achieve the property improvements that are theoretically possible for a particular combination of materials.

Interfaces constitute a research area of obvious importance. Unusual physical and chemical processes occur at the boundaries between the reinforcing phase—particles, whiskers, or fibers—and the matrix. Reactions in interfacial regions can strengthen the bonding between the matrix and the reinforcing elements, or they can weaken bonding and also degrade the constituents of the composite. Moreover, weaknesses at interfaces, due to defects or poor chemical compatibility between the matrix and reinforcing elements, are major sources of stress- or impact-induced failure of composites. One way to solve the problem of chemical incompatibility, which is an obstacle to promising combinations of materials, is to develop chemical treatments and additives to strengthen interfaces.

Composites also pose problems of geometry. The geometry of the reinforcement is a critical determinant of a composite's strength. Because fibers afford maximum control over the internal structure of a composite and because of their high aspect ratios (ratio of length to diameter), long, continuous fibers are the reinforcing elements of choice in high-performance composites (Figure 3.5). The need for more extended-chain polymers, such as the recently developed high-tensile-strength polyethylene fibers, is clear.

FIGURE 3.5   Computer-controlled 144 carrier braiding machine in the process of forming a fiberglass monocoque chassis for a composite racing car. (Courtesy Frank K. Ko, Fibrous Materials Research Center, Drexel University.)

Unless manufacturing difficulties are overcome, many promising composites, especially ceramic and metal matrix composites, may never fulfill their commercial promise. Because manufacturing involves many steps and is labor intensive, composites may be too expensive to compete with metals and polymers, even if their properties are superior. New processing technologies are needed—and not only to reduce costs of production. The high temperatures necessary for processing many ceramic and metal matrices can damage and weaken reinforcing elements. As a result, current processing technologies do not permit combinations of certain materials with especially desirable properties.

Methods for joining composites to other materials represent a research area of practical importance. Because the chemical bonding in composites is unlike that of most other materials, parts made of composites are difficult to incorporate into complex assemblies. If properties of the joining material do not equal those of the composite, the advantages of using a composite part can be nullified.

ELECTRONIC MATERIALS

Electronic materials can be semiconductors, polymers, ceramics, metals, or insulators. These materials and the processes used to fabricate them are being pushed to their limits by aggressive worldwide competition throughout the electronics industry. This has created a need for research on the fundamental limits of present technology, how these limits can be achieved by new processing methods, and what new materials or fundamental concepts will evolve to overcome these limits.

Optimization or development of individual electronic materials must be accompanied by fundamental research into how materials interact with one another. For example, information processing and communication to the external world take place through metal lines connecting semiconductor chips, logic and memory circuits, ceramic or polymer chip-carrier modules, and polymer-fiber composite cards and boards. Incompatibility between the various parts of a desired electronic component can often deter the implementation of new concepts. To enhance technology, research must take into consideration the entire materials and processing environment.

Semiconductors

Semiconductors—the source of transistors, integrated circuits, solid-state lasers, and detectors—are the central constituents of most electronic systems. Although other materials are finding an increasing range of uses, the dominant material in the electronics industry is still silicon. Among its attractive properties are its high mechanical strength, its high degree of crystalline perfection, the large diameters to which it can be grown, its natural abundance, the advantageous properties of its native oxide, and its low cost. By taking advantage of these and other capabilities of silicon, researchers have achieved high levels of circuit integration.

Competition is currently focused on dynamic memory chips, in which increased circuit density results in lower cost, higher data retrieval rates, and increased capabilities for small systems. Because devices are processed in surface arrays, lithography is of major importance, with a twofold reduction in the linear dimension of a memory cell resulting in approximately a fourfold increase in memory density. As feature sizes are further reduced, photoresist exposure to produce patterns of smaller dimensions must be accomplished with light of decreasing wavelengths. New light sources, such as synchrotron x-ray beams or deep-ultraviolet lasers, and new organic photoresist materials sensitive to the decreased wavelengths must be developed. Recent research has shown that patterns written directly with an electron beam can produce working devices with dimensions as narrow as 0.1 μm, a development that illustrates the enormous future potential of efforts to transfer new lithographic tools to industry. Control of device processing at these dimensions is equally challenging, because diffusional processes that occur with normal high-temperature fabrication techniques spread dopant profiles and reduce performance. This is causing a redirection of research toward the low-temperature processing of materials and devices, with emphasis on the thermodynamics of reactions, laser- or plasma-enhanced processing, and a more atomistic approach to understanding the behavior of materials and interactions between materials.

The advent of high-vacuum techniques for depositing epitaxial layers at low temperatures, such as molecular beam epitaxy (MBE) and chemical vapor deposition (CVD), has opened a new world of opportunity in research as well as device fabrication. Discrete thin layers can be deposited to form narrow electronic transitions or superlattices, in which multiple layers with differing compositions are alternately deposited. These techniques have been applied to silicon electronic materials. Silicon-germanium alloy layers have been grown on silicon, and devices have been fabricated whose properties depend on the difference in electronic structure between the two materials. This new flexibility will greatly extend both the range of applications of silicon technology and the ultimate performance limits of devices. Figure 3.6 illustrates strained-layer superlattices formed through epitaxial techniques whereby electronic structure is tightly controlled.

Although silicon will dominate the electronics community for the foreseeable future, much research effort is being devoted to the development of III-V semiconductor compounds such as gallium arsenide (GaAs) for specific applications. Because of their electronic structures, their high electron mobilities, and the manner in which excited electrons in III-V compounds lose their energy by emitting light, III-V materials provide higher-frequency capabilities than silicon does, and they are optically active, being both laser sources and detectors. A wide range of technologies that use these features are being investigated, such as data transmission over optical fibers, holographic image transfer, and optical storage media on disks or tapes.

One of the most important features of these materials is the number of permutations possible between elements in the group III and group V families. For example, gallium, aluminum, or indium can be combined with phosphorus, arsenic, or antimony in ternary or quaternary compounds and in multilayers produced by MBE or CVD. The ranges of laser wavelengths and device structures that can be achieved are extensive, and their limits are unknown.

Compared with elemental silicon, application of GaAs is restricted by severe processing problems related to the different chemical reactivities of

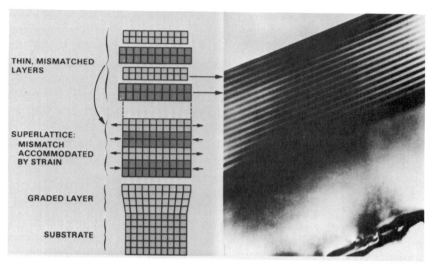

FIGURE 3.6   Strained-layer superlattices for band gap engineering. Transmission electron micrograph of a series of alternating 100-angstrom GaAs and InGaAs layers with the illustration depicting the distortion in the equilibrium lattice spacing for the two compounds that occurs in the stable structure. Stability criteria for these systems depend on the amount of lattice mismatch and the layer thicknesses, but typical cases have a 20 percent mismatch in 100-angstrom layers. Strained-layer materials are playing an increasing role in the rapidly emerging optoelectronic technologies and in a wide variety of electronic devices. By controlling the composition and thickness of the layers, researchers can optimize the electronic properties of devices. (Courtesy Sandia National Laboratories.)

the elements. Also, structural and chemical defects must be eliminated, contact fabrication is difficult, and native oxides are inadequate for use in devices. These and other materials issues must be overcome if more widespread applications of III-V compounds are to be realized.

As basic knowledge of surfaces and epitaxial layer formation has evolved, new emphasis has been given to heteroepitaxy of one material on another to achieve novel devices or to use the best features of each layer. For example, GaAs is being grown on silicon for integrated optics, in which laser or detector devices are directly associated with silicon circuitry. Vertically integrated structures containing epitaxial insulators and metals are also being studied— for example, calcium fluoride or cobalt silicide on silicon. Growth processes are difficult to control for these materials, requiring the development of new high-resolution techniques for interface studies. Until basic issues related to structural defects can be resolved, such materials will remain the subject of research. Figure 3.7 illustrates extremely small silicon structures prepared by lateral thermal oxidation processes.

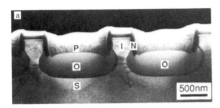

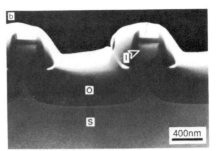

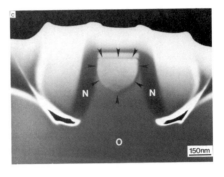

FIGURE 3.7 Artificially structured materials for microelectronics: isolated silicon islands. Future microelectronic circuits will require progressively smaller electrically isolated areas of single-crystal silicon. Panels a, b, and c show transmission electron microscope (TEM) images of *isolated* single-crystal silicon islands prepared by the lateral thermal oxidation of single-crystal silicon trench structures capped with silicon nitride. The patterns are produced by electron beam lithography, and the thin films are deposited by chemical vapor deposition processes. (a) TEM cross section showing silicon-on-insulator structures; the silicon islands have not been completely isolated. Labels indicate silicon substrate (S), silicon dioxide (O), polysilicon (P), silicon nitride (N), and silicon island (I). The dark lines are defects (dislocations) induced during processing. (b) TEM micrograph showing structures where the silicon islands have been completely isolated by encroaching oxide. Note reduction in defect density. (c) High-magnification TEM micrograph of an isolated silicon island shown in (b). The small island (outlined by arrows) is approximately 200 nm (1/500 the diameter of a human hair) on a side. (Courtesy N. C. MacDonald, School of Electrical Engineering, Cornell University.)

Ceramics and Polymers

Electronic packages, consisting mainly of wiring and interconnections, provide the interface between active semiconductor devices and the external world. The levels of packaging technology involved are related primarily to the size and to the performance requirements of the application, ranging from one or a few semiconductor chips to massively integrated, large computer systems. The requirements for a substrate material onto which chips can be attached are easily understood, but in practice these materials generate a complex set of interactions that need to be resolved. Ceramic substrates are currently used when high circuit density and multiple chips are needed and when mechanical stability is important. Substrates are fabricated by screening metal patterns onto thin, prefired ceramic layers, stacking the layers, and firing at high temperatures to densify and strengthen the mass. For alumina, which is the ceramic presently used, the choice of conductor metal is restricted to molybdenum because of the need for excessive firing temperatures. However, the dielectric constant of alumina is too high and molybdenum is too resistive, resulting in relatively poor electric characteristics for the transmission lines. Also, the thermal coefficient of expansion of alumina is not the same as that for silicon, leading to high stress and failure when chips are attached and powered.

New substrate materials must be developed that have lower fabrication temperatures, lower dielectric constants, high strength, and thermal expansion coefficients matched for those of silicon. Compatibility during processing between the substrate matrix, the metal conductors, and the semiconductor chips has been elusive and will require more research into interface interactions, fracture propagation, and thermal relaxation mechanisms. If ceramics are to remain competitive for chip substrates, dielectric constants must be reduced by techniques such as microscopic porosity, and surface roughness must be eliminated to facilitate lithographic processing of metal lines.

There is a significant trend away from ceramics and toward polymers for substrate packaging materials. Polymer layers have intrinsically lower dielectric constants and preparation temperatures, can be applied onto a surface with controllable flatness, and are compatible with lithographic processing of metal transmission lines. Mechanical stability and heat dissipation are areas requiring further research, but hybrid ceramic-polymer layers are now being produced in which the polymer layers are processed on top of or adjacent to chips when metal lines of the highest density and highest performance are required. New polymeric materials will be needed in which thermal expansion coefficients are reduced to match the chip and ceramic, strength is increased to prevent cracking, and dielectric constants are further reduced, perhaps by the use of fluorocarbon-like chemistries. Adding moisture resistance, thermal stability, and perhaps photosensitivity to the list illustrates the complex combination of requirements that apply to new polymer materials.

The printed circuit boards onto which chip-carrier modules are mounted have been made of epoxy-filled glass cloth for some time. The reinforced structure has a thermal expansion coefficient characteristic of the composite, with strength being provided by the glass. Research on this material is seeking to lower the dielectric constant by using aromatic polyamide polymers to replace the glass, and to replace the epoxy with a more compatible material. Structural integrity and compatibility with high wiring densities are further areas for research.

Many other important electronic applications for ceramics and polymers are the subjects of research, including capacitor coatings for substrates and chips, piezoelectrics, ferrite recording-head elements, bistable switching, photosensitive polymers for lithography, polymer substrates for optical storage, and high-strength polymers spun from liquid crystal solutions. Impressive advances have been made in recent years in the field of electroactive polymers. Conjugated polymers (polymers that have overlapping π orbitals along their backbone) have been reported with conductivity approaching that of copper at room temperature. Continuing work in this field is needed in processability and in performance, especially with respect to environmental stability.

Metals

Metallic components serve a number of functions in electronic systems, from transmission lines within and between packing levels to magnetooptic or magnetic films for information storage. Because high device densities in semiconductor chips cannot be achieved without providing the necessary wiring between devices or circuits on the chip, wires have to be reduced in size to the point at which issues of reliability become overriding. Wiring is also distributed vertically over many layers processed on the chip, not unlike the multilayered wiring processed on substrates, except with much finer dimensions. Research issues include the stability of the contacts made to the semiconductors; interdiffusion between, on the one hand, metal layers, contacts, and conductor lines, and, on the other hand, the structures on the chip and substrate; reliability at the high current densities resulting from finer lithography; structural and chemical compatibility with the insulating materials that separate the wiring layers; and the effects of environmental attack. New metal deposition processes using vapor transport and radiation enhancement are being developed to expand the list of candidate metals and diffusion barriers that can be implemented. Specific metallurgy questions may vary between silicon and GaAs, but the fundamental scientific issues are similar. Figure 3.8 shows multilayer ceramic packages with metallic conductor lines deposited on the ceramic substrate providing complex interconnections among active devices.

Developing new materials and processes and performing extraordinary research and laboratory work are necessary but are not sufficient to ensure the competitiveness of the electronics industry. Engineering of needed advances into manufactured products is one of the biggest challenges facing the materials community. Those companies that are proficient at this step of the innovation process will be leading edge competitors.

FIGURE 3.8   Multilayer ceramic packages. (Courtesy IBM Corporation.)

MAGNETIC MATERIALS

Magnetic materials have received less public attention during recent decades than have semiconducting or superconducting materials. Nevertheless, the intrinsic scientific issues and opportunities in the study of magnetic materials are equally interesting, and the existing and potential economic impacts of magnetic materials are comparably great.

The economic impact is perhaps easiest to see. Magnetic materials underlie applications in information storage, audio and video recording, microwave communications, and permanent magnets for products ranging from motors to wall clips. Of these applications, the largest by far is the magnetic storage industry. The entire field of data processing stands on two legs—semiconductor logic and memory to process data, and magnetic devices to store data. Not surprisingly, the total yearly sales of magnetic storage systems, totaling several tens of billions of dollars, are comparable to those for the semiconductor logic and memory systems in computers.

For years, the magnetic storage industry has used standard materials— iron and chromium oxides for storage media and permalloy or ferrites for heads. The bottleneck to increasing the size and efficiency of these systems was never the magnetic materials but rather the mechanical components required to fly recording heads faster and closer to the magnetic medium. But now the magnetic materials themselves are beginning to limit further progress, because of recent developments such as thin-film heads; the demand for higher-density storage, requiring new storage media; and the emergence of competitive new technologies, such as magnetooptic storage, that call for entirely new classes of magnetic materials. As never before, magnetic materials are the key to the future of the storage industry.

Scientific opportunities in magnetic materials, exciting in themselves, also dovetail in many cases with the technical demands of magnetic applications. An example is the study of magnetic surfaces and interfaces. Control and study of surfaces on an atomic level have proved to be more challenging for magnetic materials than for semiconducting materials. One reason is that detection of magnetic properties requires the use of so-called spin-sensitive measurement techniques, which add extra complexity to the already very sophisticated surface analytical tools used to study semiconductors and other nonmagnetic materials. However, this technology is now largely in place in many laboratories throughout the world, opening the door to an understanding of magnetic surfaces that was never before possible.

Magnetic surfaces and interfaces can give rise to anisotropies and effective fields that become an ever larger factor in the behavior of the bulk material as its thickness decreases, which will be a major consideration in future storage technologies. A specific example concerns the peculiar internal field, called exchange anisotropy, that arises at interfaces between a ferromagnet and an antiferromagnet, causing the ferromagnet to behave as if it were subject to an external field. Such an effect could potentially find application where it is awkward or costly to apply a field with an external power source. This long-puzzling phenomenon is just beginning to be understood in terms of the randomness of the interfacial magnetic structure. Synthesis of more perfect interfaces would offer a conclusive test of the mechanism.

Control of magnetic surfaces and interfaces is also the basis for synthesis of new artificial magnetic materials—multilayers and superlattices. Early experiments have already shown complex couplings between the different magnetic layers, which could be the basis for new properties not available in bulk materials. There is hope for achieving larger anisotropies, coercivities, galvanomagnetic effects, and magnetooptic effects, and there are tremendous varieties of systems to explore. Interest in this area has grown dramatically in the last few years in research groups throughout the world.

Another example of scientific opportunities comes from progress in the band theory of magnetic materials. Using novel statistical techniques and Monte Carlo calculations, researchers are now calculating the Curie temperature of iron with increasing accuracy. This had been a long-standing problem, and its solution opens the door to theoretical prediction of a variety of other magnetic properties. Many of these fall under a common umbrella, the so-called spin-orbit interaction, which gives rise to anisotropy, galvanomagnetism, and magnetooptic rotations. All of these properties have been poorly understood in the past and are decisive for many applications.

Major advances in magnetic materials have been appearing at an accelerating pace. New neodymium-iron-boron materials have been discovered that significantly increase the energy product of permanent magnets, thus increasing the volumetric efficiency in applications such as motors. Record magnetooptic rotations, larger by a factor of 20 than those presently used in magnetooptic systems under development, have been discovered in a new class of uranium chalcogenides. During the last decade, a whole new class of magnetic materials has been identified, called spin glasses, in which random magnetic interactions lead to complex magnetic behavior unlike any known in standard ferromagnets, ferrimagnets, and antiferromagnets. The theory of these systems constitutes a major revolution in statistical mechanics, introducing new and sophisticated concepts like distributed order parameters and nonergodic phase transitions. The analogy of random spin glass interactions and the seemingly random neuronal couplings in the brain have led to the concept of neural nets, resulting in a new theoretical approach to distributed data storage that is now being actively explored in the computer science community.

These and many other remarkable developments point to a remarkable richness of research opportunities and a large impact on applications in the field of magnetic materials.

PHOTONIC MATERIALS

Photonic materials are now where electronic materials were in the early 1950s—at the very beginning of a steep growth curve. In information technology, a photonic or optical material is any material with which light interacts for the purposes of information generation, transmission, detection, conversion, display, storage, or processing. Like electronic technology, optical technology ultimately depends on the discovery, development, and manufacture of reliable and economical materials. The past 20 years have brought unprecedented advances across all fronts. A few of the most important are briefly mentioned below.

Before the early 1970s, the most transparent glass that could be manufactured had an optical attenuation of over 1000 dB/km. Today, optical fibers are being manufactured with loss coefficients nearly 10,000 times lower. This advance has transformed the field of optical communications from a visionary dream to a commercial reality in just over a decade.

Other important advances include the invention of room-temperature semiconductor lasers and the development of ultrasensitive, low-noise detectors based on III-V semiconductor materials. The bit rate × distance product, a figure of merit for optical communications systems, increased a thousandfold in a single decade.

Epitaxial deposition of films has led to the generation of totally new semiconductor devices, called multiquantum wells, that exhibit phenomena not found in any natural material. This discovery makes possible virtually unlimited flexibility in the design of integrated optical and optoelectronic circuits.

A whole range of new nonlinear phenomena has been discovered, and an active search is now under way for materials that optimize these phenomena for specific applications. Optical bistability provides the key to all-photonic switching and logic and quite possibly will provide the basis for an all-optical computer. Self-induced transparency and a host of related nonlinear optical propagation phenomena have recently been demonstrated in rare-earth-doped optical fibers. The photorefractive effect is a form of dynamic hologram and may form the basis for optical storage, processing, and display technologies of the future. Only a small fraction of the materials exhibiting these effects have been investigated, and the basic physics is not well understood in all cases.

Advances that bear on the display of information include the intrinsic photo-ferroelectric effect, which may be used to store high-quality images in transparent ferroelectric ceramic materials. In addition, the discovery of ferroelectric liquid crystals has led to new electrooptical devices that respond extremely rapidly to external electrical signals.

Although this section focuses on applications related to information technologies, optical materials also have uses in many other areas that are being actively investigated. Recent materials research has improved the conversion efficiencies and lowered the production costs of solar cells made from amorphous silicon, bringing this technology to the threshold of large-scale commercial viability. Ultraviolet lasers are being developed for laser fusion, while other lasers are producing very short pulses of light that can be used to study chemical reactions and other dynamic processes.

Substantial advances have been made in many other areas of photonics, such as optical phase conjugation, two-wave mixing and energy exchange, erasable optical memories, high-power optical materials, gradient-index coatings, phase transition materials, flat panel displays, and optical window materials. The development of suitable high-performance materials is the key to success in all these areas.

Virtually all of the discoveries discussed above suggest directions for further research. Listed below are a few of the obvious opportunities within the major categories.

For fibers or other transmission systems, it is important to devise new materials that have lower optical losses. Probably required are transparency windows at wavelengths longer than 1.5 μm. For use in power transmission or medical applications such as surgery, a modest-loss glass operating with a transparency window near 10 μm is important.

Important possibilities exist for the use of fibers in nonlinear optical processing—that is, in the regeneration, amplification, detection, and modulation of pulsed signals. Such applications require active glasses such as silica doped with rare-earth ions.

The theory of the nonlinear propagation of novel optical modes (e.g., solitons), the interactions between these modes, and the optimization of waveguides designed for these nonlinear effects deserve increased attention.

If the speed advantage offered by photonics over electronics is ever to be matched at least by parity in the spatial domain, materials or structures must be found that permit optical modulation over micron rather that millimeter distances. Increasing the nonlinear optical response of photonic materials is thus a fundamental challenge to materials scientists and theoretical physicists.

The search for optical silicon, although difficult, is nevertheless important. Optoelectronic semiconductor materials, in addition to the group IV and III-V systems, require increased attention; II-VIs and I-III-Vs, both in bulk and superlattice forms, provide intriguing possibilities.

In the materials science of crystal growth, European and Japanese research efforts dominate U.S. efforts. There is a growing need for bulk crystals of various kinds as well as for specific magnetooptic, acoustooptic, and electrooptic materials. The scarcity of such materials often limits the scope and pace of research on new phenomena, devices, and systems. Figure 3.9 illustrates growth of high-purity lithium niobate crystals for electrooptic applications in telecommunications.

Another challenge of great significance is the extension to two and even three dimensions of the kind of control in materials fabrication that vapor deposition provides in one dimension. The extension from quantum wells to quantum wires and then to quantum dots is but one example of the power that such an advance in fabrication technology would represent.

Serious consideration should be given to a research program aimed at what might be called microprogrammable materials. The phenomena of spectral hole burning and photorefractive effects represent limited examples of this idea. It might be possible to use energetic beams to make localized changes in the chemical bonds or the structure of a material in a way that will change its function at desired points in space. For example, it might be possible to convert an insulator into a semiconductor or a semiconductor into a superconductor in spatially controlled regions.

Finally, as materials become purer and more microscopically controlled, the requirements for materials characterization become even more stringent. Monitoring the amount, identity, and location of each chemical or even each atomic constituent in situ in real time, and often during the fabrication process itself, will require substantial research in characterization. The tools and technological sophistication required are complex, and the costs of the equipment are high. Nevertheless, characterization is essential to successful materials developments, and the investment should be made.

FIGURE 3.9   View of Czochralski crystal growth furnace preparing lithium niobate. Czochralski growth in which a seed is dipped into molten material contained in a crucible and slowly extracted is the principal technique for the preparation of silicon, III-V semiconductors, and refractory materials used for lasers and electrooptic modulators. Lithium niobate is an electrooptic material used for switching light signals at high speed. The melt is at a temperature above 1000°C. (Courtesy AT&T Bell Laboratories.)

SUPERCONDUCTING MATERIALS

The recent discovery of high-temperature superconductivity in ceramic oxides is one of the most dramatic breakthroughs in physics and materials science in recent decades. It has focused world attention on the field of superconductivity; more broadly, it has quickly come to symbolize the opportunity and challenge in the overall field of materials science and engineering.

Nevertheless, high-temperature superconductivity is only one part of the story of superconducting materials. Throughout the decades following H. Kamerlingh Onnes's discovery, in 1911, of superconductivity in mercury at 4.2 K, materials with ever higher transition temperatures were steadily discovered, until 23 K in an intermetallic compound of niobium-germanium was reached in 1973. At the same time, theories developed in the middle to late 1950s and early 1960s provided a comprehensive understanding of the existing experiments based on the formation of electron pairs mediated by lattice vibrations. With the development of superconducting magnetic materials such as niobium-titanium and niobium-tin, which can carry large currents in the presence of large magnetic fields and remain superconducting, and of Josephson devices such as superconducting quantum interference devices (SQUIDs), sales of complete systems based on superconducting components have approached $1 billion annually.

In spite of the long hiatus in increasing transition temperatures beyond 23 K, the field of superconducting materials has not remained static. In fact, since 1973 a series of novel materials and systems have been discovered and studied. Magnetic superconductors have been found in which superconductivity and magnetism coexist (normally, superconducting materials expel magnetic fields), superconductors containing organic materials have been developed, artificially structured arrays have been shown to exhibit complex behavior known as frustration, ultrasmall junctions have displayed novel quantum behavior not seen in standard Josephson junctions, and heavy-fermion superconductors have demonstrated novel types of pairing. These studies have raised fundamental issues about the nature of the microscopic pairing mechanisms in highly correlated electron systems, the effects of disorder, and the underlying quantum description of dissipative systems.

These superconducting materials, which so far have shown low to modest transition temperatures, have not captured the public or wider scientific imagination. However, their underlying physics and the materials issues associated with them are of great complexity and interest. At present, the questions they raise are by no means resolved and deserve further attention. And the possibility for setting new transition temperature records should not be discounted: there has been steady progress, for example, in raising the transition temperatures of organic superconductors to a level approaching that of "pre-Bednorz-Muller" oxides.

Building on the earlier discovery of superconductivity at up to about 13 K in certain oxides, J. Georg Bednorz and Karl Alex Muller in early 1986 found superconductivity at over 30 K in a class of ceramic oxides containing lanthanum, barium, and copper. Shortly after this result was verified in December 1986, superconductivity at up to about 95 K was discovered in a compound of yttrium-barium-copper-oxygen with a structure now dubbed "123."

More recently, zero resistance at up to 107 K has been reported in layered structures of bismuth-strontium-calcium-copper-oxygen. Related structures of thallium-barium-calcium-copper-oxygen have pushed the zero-resistance point to 125 K. It is noteworthy that the key breakthroughs in new materials have come both from small laboratories and from larger centers in different countries, demonstrating the importance of a balanced national and international research effort.

There have been many reports of higher transition temperatures. So far these results have been difficult to reproduce, and there is no consensus at present about their validity. Nevertheless, there is widespread expectation that new phases will be discovered, and a frenetic search for them continues in many laboratories worldwide.

Synthesizing and characterizing these new materials have been multidisciplinary enterprises involving contributions from many fields. Unlike most earlier superconductors, the new materials are ceramics, so that many techniques of ceramic processing have been applied to their synthesis. The new superconductors have already been prepared in the form of wires, thin films, and single crystals, in addition to sintered granular material. Yet many challenges remain: the ceramic or polycrystalline material suffers from low critical current densities (the highest current the material can carry before losing its superconductivity), and the single crystals are still generally too small for decisive neutron-scattering studies, which would allow their magnetic correlations to be determined. Electron pairing also has yet to be controlled in single-crystal synthesis.

The problems in synthesis have also pointed to a variety of chemistry issues, including the interactions of the new high-temperature superconductors with environmental factors such as water and carbon dioxide. The large number of elemental constituents and the critical role of oxygen in the structures make their chemistry both complex and interesting. Formation of the superconducting phase seems to require high-temperature annealing, but proper control of the oxygen, at least in the "123" compounds, requires a lower temperature. Whether these temperatures can be lowered further will be vital to many applications that cannot tolerate high-temperature anneals. This is particularly true of thin-film applications, and studies have already revealed significant interdiffusion of the superconducting films with most substrates.

The nature of superconductivity in these materials and the nature of the materials in their nonsuperconducting states have rapidly become central problems in modern condensed-matter physics. There is growing evidence that new pairing mechanisms are required that go beyond the conventional electron-phonon coupling believed to account for most earlier superconductors. There is also growing evidence that earlier theories must be generalized to include effects arising from both large anisotropy and fluctuation phenomena enhanced by the short coherence length in these materials. The effects of different oxygen defect structures, of twin boundaries, and of copper oxide planes and chains on the superconductivity are all yet to be understood. The structure of a superconducting material with one of the highest transition temperatures yet achieved is shown in Figure 3.10.

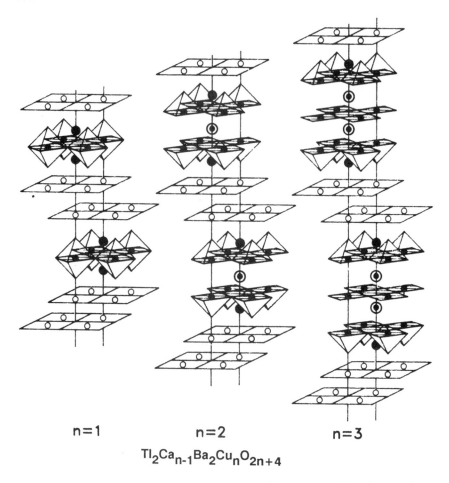

$$Tl_2Ca_{n-1}Ba_2Cu_nO_{2n+4}$$

FIGURE 3.10   Structure of a superconducting material. As the number of copper layers (n) increases from one to three, the superconducting transition temperature rises. (Courtesy Stuart Parkin, IBM Corporation.)

A vast number of experiments have been conducted to elucidate the microscopic mechanisms of the new superconductors. But basic information on high-quality single crystals has only begun to appear, and vital experiments such as neutron scattering still await the availability of large enough crystals. Although many theories have been proposed, most theorists agree that more experimental information is needed before a consensus will begin to emerge.

Many other physically interesting properties bear on the practical application of these materials. Foremost is the critical current density, which appears to be controlled by the presence of grain boundaries in polycrystalline material. The nature of those grain boundaries and the control of impurity phases segregated at them are just beginning to be studied. In single crystals, the nature of the mechanism that permits large current densities is as yet unknown. There has been steady but largely empirical progress in increasing current densities for practical application. More work is needed before such applications as magnetic wire will become feasible.

There are numerous structural issues as well. In ceramics, various superconducting structures have already been synthesized, with different degrees of orientation. Again, this work has been done on a largely empirical basis, but a detailed understanding is vital for control of mechanical and electrical properties. In thin films, epitaxial growth has been used to demonstrate the high critical currents possible in these materials. These findings need to be demonstrated in polymeric films.

BIOMATERIALS

Biomaterials can be broadly defined as the class of materials suitable for biomedical applications. They may be synthetically derived from nonbiological or even inorganic materials, or they may originate in living tissues. The products that incorporate biomaterials are extremely varied and include artificial organs; biochemical sensors; vascular grafts; artificial hearts; disposable materials and commodities; drug-delivery systems; hybrid artificial organs; dental, plastic surgery, ear, and ophthalmological devices; orthopedic replacements; prostheses and repairs; wound management aids; and packaging materials for biomedical and hygienic uses.

The success of already developed materials, notably of implants and joint replacements in older patients, has helped create an increased demand for these devices in all age groups. A corollary of this new demand is that the in vivo lifetime of devices will have to be increased from a current value of approximately 20 years to possibly 80 years.

The search for new biomaterials has accelerated enormously in the past two decades, spurred on in part by rapid advances in biomedical research and by the development of new synthetic materials, particularly polymers. A primary direction of biomaterials research must be toward understanding the interaction between synthetic substrates and biological tissues to meet the needs of clinical applications. Much basic research is currently directed toward finding new materials for reconstructive surgery and for other applications. However, the anticipated rapid increase in the demand for biomaterials will require that applied research also focus on cutting the costs of substance manufacture, clinical procedures, and rehabilitative systems.

Biomaterials Categories

Synthetic Nondegradable Polymers

One of the most important research issues related to synthetic nondegradable polymers is their stability and compatibility in the presence of biological materials, since they are often used in applications requiring moderate to long lifetimes. Their very broad range of applications includes prosthetic devices, vascular grafts, sutures, packaging, soft tissue augmentation, drug-delivery systems, synthetic oxygen carriers, contact lenses, and orthopedic replacements, as well as packaging, syringes, and diapers. For each material, stability and compatibility issues include compatibility with tissues (including blood), hydrolytic stability, reactivity with drugs and medications, calcification, long-term functionality, mutagenic or carcinogenic properties, and sterilizability.

Concentrated efforts should be directed to investigating in vivo reactions with tissues, the effects of surface chemistry and morphology on thrombogenesis, protein and cell consumption, and the formation of emboli. Additional concerns include assessing the medical risks of additives in materials, low-molecular-weight components, in vivo degradation products, and sterilization products, and constructing a data base to help assess these biomedical factors.

Synthetic materials used in drug-delivery systems should be compatible with a number of different drugs and other excipients and should be stable during processing and during sterilization by gamma radiation.

Bioabsorbable and Soluble Polymers

Unlike the nondegradable polymers, bioabsorbable compounds depend for their functionality on a capacity to be absorbed by living tissue and/or dispersed in a biological solvent such as blood serum. Applications for this class of synthetic biomaterials include sutures, vascular grafts, artificial ligament scaffolds, drug-delivery systems, environmentally degradable materials, and edible food casings. Figure 3.11 illustrates the growth of endothelial cells on a vascular graft.

Some of the key issues for these materials are absorbability and its measurement and definition, the effects of absorption on the tissue site, the effects of enzymes and other biologically active materials on the absorbability of the polymers, the biological effects of degradation products, the effects of sterilization on functionality and degradability, the behavior of labile release agents incorporated into the polymer, and the effects of substances on wound healing.

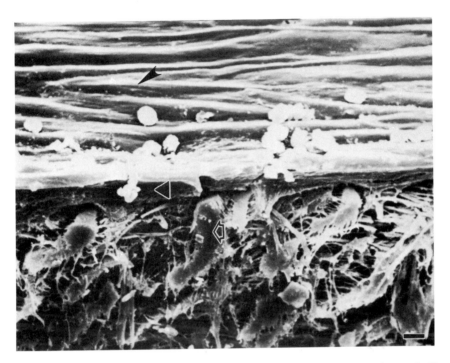

FIGURE 3.11 Scanning electron microscope micrograph of a vascular graft "seeded" with autologous microvessel endothelial cells and implanted in a canine artery for 5 weeks. The arrowhead points to the endothelial cell lining that was created. The triangle shows the thickness of the new cellular lining. The open arrow points to the vascular graft. Bar = 11 μm. (Reprinted, by permission, from Bruce Jarrell and Stuart Williams. Copyright © 1989 by Bruce Jarrell and Stuart Williams.)

Polymeric Tissue Adhesives

The strength and durability of surfaces attached by polymers are clearly of essential importance, as are the chemical and mechanical compatibility of tissue adhesives with biological materials. The problem of tissue-to-tissue adhesion will require careful study. In addition, research will be required to enable understanding of the biological effects of degradation products of adhesives, the biological pathways that interact with degradation products and the effects of degradation products on those pathways, the effects of processing and sterilization on biodegradation, and the effects and control of wound healing.

Metals, Ceramics, and Glass

Metals and metal alloys are broadly used in biomedical applications, such as in orthopedic replacements, plastic surgery, and sutures. For metals to meet criteria for biomedical use, they must be well characterized with regard to electrochemical corrodibility, the durability of passive layers, other degradative mechanisms, ability to elicit immune and hypersensitive reactions, and interactions with biological pathways.

For ceramics and glassy materials, durability, fracture toughness, surface activity, and degradation are properties requiring in-depth investigation. The biological effects of degradation, the effects of biological materials such as enzymes, and the effects of processing and sterilization on the behavior of synthetic biomaterials are also important issues.

For bioabsorbable ceramics, which are projected for use in orthopedics, research should focus on the measurement of bioabsorption; the influence of particular tissue sites; the effects of the ceramics on calcification, bone formation, and wound healing; and the ramifications of incorporating labile release agents into the ceramics.

Fundamental research on bioactive ceramics should be directed to bioactive mechanisms. Specific issues include the compositional dependence of bioactive bonding to bone and soft tissues, the effects of mismatches of elasticity modulus on stress transfer, bioelectric potentials and interfacial bonding properties, and the stability and fatigue resistance of bioactive interfacial bonding, coatings, and composites.

Carbons

Carbon in various forms—glassy, pyrolytically deposited, in fibers, and in composites—is a candidate material for biomedical applications. Like other proposed synthetic biomaterials, carbons must be characterized regarding properties affecting biological activity. The strength and fracture resistance of carbon materials are salient properties for biological use.

Biologically Derived Materials

Biological materials from both human and nonhuman sources have medical applications. Such materials include processed tissue from porcine heart valves, bovine carotid arteries, human umbilical veins, reconstituted collagen and elastin, hyaluronic acid, chondroitin sulfate, and chitosans. Applications include cardiovascular and ophthalmic applications as well as soft tissue augmentation.

Whether human or nonhuman, biological materials tend to provoke antigenic rejection responses, which need to be characterized. The materials noted require isolation and purification, and standards for purity must be defined. Durability of materials and their calcification effects are also important issues.

Surface Modifications

Numerous techniques of coating and surface derivatization are under investigation in a variety of medical applications, including the control of tissue adhesion, the reduction of friction (e.g., in catheters), thrombolytic and anticoagulant agents, antibiotics, immobilized enzymes or other agents for chemotherapy, and removal of certain proteins or cells. Central issues for each of these processes must be the complete biological characterization of the coatings or surfaces, measurement and definition of in vitro and in vivo biological activity, and the effects of sterilization on the functionality of the coatings.

Major progress in extracorporeal procedures could be achieved if device surfaces could be rendered nonthrombogenic at reasonable cost. The most significant barrier to success at the moment is lack of an appropriate model for human use, as platelet aggregation and the clotting cascade mechanism appear to be unique to humans; hence adequate animal models are not available.

Surface effects are also crucial to determining immunoactive or immunocompatible properties. Creation of surfaces with affinities for specific proteins will probably be a major focus for the development of new polymers.

Long-Term Opportunities Related to Biomaterials

The incorporation of biochemical and pharmacological activity into medical devices is an area that is already undergoing rapid development. Particular areas of interest include devices for controlled and sustained drug delivery and devices with active anticoagulant behavior and antibacterial properties.

Totally synthetic or artificial devices will not be used for very long in medicine. The field is now moving toward use of what can be called hybrid devices and organs. A good example is the use of materials as a scaffold on which to culture, organize, and grow cells for specific functions. This is happening, for example, in the creation of the artificial pancreas. Today's vascular grafts and artificial blood vessels depend, in part, on cellular linings for their function.

The biotechnology industry also depends on the growing or culturing of cells on suitable supports for various purposes. This field is undergoing rapid development, due largely to the discovery and increased understanding of cellular growth factors, which allow cell growth and proliferation to be controlled.

An extension of hybrid devices and organs will be the capacity to induce cells and tissues to fully duplicate and regenerate themselves. There is some evidence for this in nonmammalian species. Developing a means to literally grow a new heart or kidney from a piece of biopsy tissue would be far more desirable and perhaps less costly than attempting to maintain an individual on an artificial kidney or heart indefinitely.

With regard to basic research opportunities in biomaterials, a number of recent developments in surface chemistry and surface characterization, as well as in microengineering and microprocessing, can be expected to lead to major breakthroughs and innovations in the near future. Specific examples are the scanning tunneling and scanning force microscopes, which provide for the study, analysis, and understanding of surfaces at the subnanometer scale. These devices are only beginning to be applied to solids in an aqueous environment and to biological macromolecules, including proteins and viruses. In principle, they permit imaging of individual atoms on the surface of, for example, protein molecules deposited on an appropriate support.

Of particular interest and potential application is the scanning force microscope, which images intermolecular and interatomic forces of attraction with subnanometer lateral resolution. If such an instrument can indeed be applied to biological macromolecules, viruses, and cell surfaces, it will transform research and understanding in these areas.

There has been considerable progress in adapting micron- and submicron-scale processes commonly used in the electronics industry to biological materials. Thin-film and integrated optics, micromachining, and microarea chemistry and biochemistry are all being applied extensively to the development of biosensors, including miniature multichannel biosensors, which eventually will be incorporated into medical devices. This will permit regulation of function and will enable the multivariate testing and evaluation of medical devices in a variety of complex biological environments.

Microengineering processes are already being used in the cell culture field to deposit cells on appropriate substrates in microscopic arrays and patterns. As this technology evolves, it will be possible to envision cells of the same or different types being deposited and maintained in precisely designed morphologies and complex three-dimensional architectures. One could, in principle, begin to assemble synthetic tissues, using individual cellular and macromolecular building blocks.

FINDINGS

Over the past 10 decades, enormous improvements in materials and in their ability to perform useful functions for society have changed the world we live in—our transportation, communications, health, safety, and quality of life. One might reasonably ask, are we nearing an end to all this? Are we approaching a leveling off, a period of diminishing returns? The findings of this committee, to the contrary, are that the ripeness of the field is undiminished and that new materials and functional improvements of materials will continue to come at a remarkable rate. Of course, that rate will depend on resources devoted to the task. Those nations that deploy their resources wisely can be expected to reap a rich technological and economic harvest in the decades ahead.

Several common themes have been discussed in this chapter. Foremost is the role of synthesis and processing in developing structural, electronic, magnetic, photonic, and superconducting materials and biomaterials. From strip casting of metals through the synthesizing of new nonlinear optical media in photonic materials to improving the critical current requirements of the new high-temperature superconductors, synthesis and processing are crucial for the continuing advancement of those technologies that depend on these functional properties.

Numerical simulations, modeling, and calculation of properties from first principles are increasingly used by scientists and engineers as means to shorten the time to develop understanding and applications of materials. It is almost inevitable that computers will be used increasingly to simulate the properties of new assemblies of atoms or to simulate a new process to shape structural materials, for instance. This trend needs to be encouraged in materials science and engineering both at the level of practice as well as at the level of teaching.

The enormous utility of the functional properties of the materials described in this chapter is well known. Previous studies dealing with particular classes of functional materials, such as electronic materials, have been carried out both in the United States and abroad. This study, however, emphasizes to a much greater degree the importance of synthesis and processing as the key, top-priority area of opportunity in materials science and engineering today. This theme is further developed in the section "Synthesis and Processing" in Chapter 4. To a very large extent, the relative advantage an organization or a nation will have in the competitive race ahead will be determined by the speed with which ideas such as those presented here can be exploited.

4

Research Opportunities and the Elements of Materials Science and Engineering

In Chapter 3, research opportunities in materials science and engineering were described in terms of the functional roles of materials. Such a breakdown has the advantage of making explicit the link between fundamental research and its applications. Inevitably, however, a presentation organized in that way omits some important areas of research. Often in basic research, the focus is structure, properties, phenomena, or behavior of materials; the functional utility of a new material may not be appreciated until its properties have been adequately characterized. One example is quasi-crystals, a class of materials neither exactly crystalline nor amorphous but with some attributes of both. These materials were discovered only in this decade. Properties of this class of materials are still being measured, and it is not clear which functional property will result in an application. Other research topics may be linked to applications but may still not be adequately described by a breakdown of research opportunities according to materials function, because such topics may relate broadly to several, or all, of the functional classes of materials. An example is rapid solidification processing, which has already had important applications in metals for both structural and magnetic applications, and in ceramics for structural electronic applications.

Science is a process of collecting and organizing knowledge about nature. The motivation of the scientist may vary widely. In some cases, it is to understand a phenomenon at its most fundamental level. In others, the motivation is to understand and gain knowledge in a particular area that is believed to be useful in developing materials for a particular application. The latter motivation is easy to justify because it aims at immediate results. What is less easy to explain is the pursuit of science when application is not the primary motivation. In physics and chemistry, the value of seeking knowledge for its own sake tends to be taken for granted. In the field of materials science and engineering, the role of fundamental science is just as essential and forms the basis for many of the great developments that have already changed our society and will continue to transform it.

For example, quantum mechanics is the basis of our understanding of solids. The use of quantum mechanical principles in building transistors or future quantum-well devices is a case in point. Without the framework provided by this fundamental knowledge, the power of materials science and engineering, which led to the integrated circuit, would never have revolutionized information technology. The importance of quantum mechanics appeared again in the invention of the first and of all subsequent lasers. It is now playing a central role in the development of high-T_c superconductors. The discovery of high-speed superconducting Josephson junctions was not motivated by applications. But their development into a high-speed switch (to which materials science and engineering made a contribution) was so motivated. The new high-temperature superconductors were not discovered during a search for applications. But their development as useful materials will require scientific research directed toward applications.

The point is that fundamental scientific advances are essential components of materials science and engineering, even though the motivation for them may, at the outset, have no obvious connection with applications. Once a major scientific breakthrough occurs, the full power of materials science and engineering is needed to make something useful of it.

The methodology for developing materials for applications, which provides an underlying coherence to this diverse field, is the framework for this chapter's examination of research opportunities in materials science and engineering. Assessing research opportunities in terms of the four basic elements of the field (see Figure 1.10) allows study of their relative significance regardless of materials class, functional applications, or position in the spectrum from basic research to engineering. It also provides another perspective from which to discern and monitor broad trends in materials science and engineering. In this chapter, emphasis is given to synthesis and processing as the element of the field that especially requires the concentrated efforts of U.S. researchers.

In the past, many viewed materials science and engineering as focusing primarily on structure-property-performance relationships in materials. This view of the field has had its counterpart in the activities viewed as important by materials scientists and engineers. Materials scientists and engineers have studied the structure and composition of materials on scales ranging from the electronic and atomic through the microscopic to the macroscopic. They have measured materials properties of all kinds, such as mechanical strength, optical reflectivity, and electrical conductivity. They have predicted and evaluated the performance of materials as structural or functional elements in engineering systems.

In conducting all these studies, materials scientists and engineers have recognized that properties and performance depend on structure and composition, which in turn are the result of the synthesis or processing of a material. Only recently, however, have synthesis and processing come to be widely viewed as an essential and integral element of materials science and engineering. As discussed in Chapter 3, synthesis of new materials by unusual chemical routes and by various physical and chemical means has led to an era in which atom-by-atom fabrication can be achieved. Coincidentally, processing has received renewed attention, partly in response to challenges from international competitors who have reaped the benefits of improved quality and uniformity of traditional materials, and partly in response to the demands for process control to achieve the promise of advanced materials.

Increasingly, work in materials science and engineering involves interactions among groups working in all elements of the field. One example of this trend is the development of new high-temperature intermetallic composites to achieve a complex array of performance-driven properties for applications such as the proposed national aerospace plane. If materials science and engineering is to remain healthy and productive, research on all elements of the field—and on their relationships—is vital. Nonetheless, the committee has emphasized synthesis and processing as the aspect of the field representing the greatest national weakness and also the ripest opportunities. In addition, concluding sections of this chapter discuss instrumentation and modeling, which are the areas of research critical to synthesis and processing as well as to the other elements of the field.

More detailed discussions of these significant aspects of materials science and engineering are presented in Appendixes A, B, C, D, and E, which describe important research opportunities in synthesis, processing, performance, instrumentation, and analysis and modeling, respectively. This information should be useful to the practitioners of materials science and engineering as well as to the federal agencies seeking specific advice on technical areas of research.

PROPERTIES AND PERFORMANCE

Properties are the descriptors that define the functional attributes and utility of materials. The brilliance and transparency of diamond, for example, give rise to its use as gemstones as well as sophisticated optical coatings, while its great hardness and thermal conductivity permit quite different applications such as cutting tools and media. A micrograph of a diamond thin film is shown in Figure 4.1. Metals are ductile, a property that facilitates their being processed into wires for electrical conduction or for mechanical retention.

FIGURE 4.1   Polycrystalline diamond film deposited on silicon wafer by the hot filament chemical vapor deposition method. The largest triangular face is approximately 2 μm on a side. (Courtesy National Institute of Standards and Technology.)

Ceramics have high melting points, great strength, and chemical inertness that promote their use as liners or protective coatings in advanced heat engines; however, their lack of plasticity is a detrimental property that currently limits the widespread application of ceramics. Unique physical properties of polymers make possible diverse products such as sonar devices, liquid-crystal displays, electronic package encapsulation, and automobile interiors, but, conversely, the transport properties of polymers' constituents accelerate degradation, as illustrated by the ''new car'' smell present in newly assembled vehicles.

In the broadest sense, materials properties represent the collective responses of materials to external stimuli; for instance, electrical or thermal conductivity are the measured result of the application of an electric field or temperature gradient. Similarly, magnetic susceptibility, superconducting transition temperature, optical absorption, mechanical strength, elastic constants, and the like are responses to other stress fields. Collectively, properties are the quantitative measures of the electrical, magnetic, optical, thermal, and mechanical character of materials and result from the structure and composition of the synthesized or processed substance, be it in solid, liquid, or gas form or in the microscopic or macroscopic size regime.

Property measurements and their analysis are the domains of theoretical and experimental physicists, physical chemists, metallurgists, ceramists, polymer chemists, and engineers of all fields. Characterizing of some properties on an atomic scale is a research area at the forefront of theoretical physics—the theories of phase transitions in magnetic materials and in ferroelectrics are cases in point. Similarly, elucidation of the fracture mechanics of brittle ceramic and metallic materials is a focus of research by metallurgists, ceramists, and solid-state physicists alike. The understanding of properties and structure in tandem has enabled synthetic chemists to make materials with "better," or at least predictably different, properties. This interrelationship affects the ultimate utility of materials—their performance. Two of the three Nobel Prizes cited in Chapter 1—for discovery of the quantum Hall effect and for work on high-T_c superconductivity—were awarded for research based on innovative property determinations. The work on high-T_c superconductivity, of course, was also a triumph of synthesis and analysis.

With increased understanding of the origin of properties in the structure and composition of materials has come the opportunity to design desired combinations of properties. For example, in years past, engineers were limited by materials that were either strong or tough, but rarely both. Today these properties can be achieved in appropriate balance in a variety of materials, from high-strength steels to polymer composites to toughened ceramics. These new materials are often slowly introduced into applications due to the need for engineering design data that have been evaluated for accuracy and are available in convenient computer format. The need for such materials property data for design is universally recognized but is rarely given adequate attention. Collaboration among universities, government laboratories, and, most importantly, industries (in which most of the data are first generated) is critical to the development of such evaluated data bases.

Performance is the element in which the inherent properties of a material link up with product design, engineering capabilities, and human needs. The properties of a material are put to use to achieve desired performance in a device, component, or machine. Examples of measures of performance include lifetime, speed (of a device or vehicle), energy efficiency (of a machine or current carrier), safety, and life cycle costs.

Materials researchers who are concerned with performance seek to develop models that relate device performance to the fundamental properties of the component materials. They also seek to understand how the materials properties are affected in service and how to predict and improve these changes in properties. The working environment is usually highly complex, involving multiple, often synergistic stimuli and forces, such as heat and mechanical stress cycling, moisture and oxidation exposure, and irradiation. For structural materials, for example, it is necessary to know how the materials respond to stresses caused by service loading, mechanical contacts, or temperature variation; how they react to corrosive or otherwise hostile environments; and how they undergo internal degradation. The crucial issues are reliability, durability, life prediction, and life extension at minimum cost. Understanding of failure modes and development of rational simulative test procedures are crucial to the development of improved materials, designs, and processes. Such issues are relevant not only for materials used in large structures or machines, but also for those that form structural and other elements in electronic, magnetic, or optical devices.

Behavior in service generally refers to the behavior of a material in some end use—as in a turbine blade, piston head, containment vessel, energy-absorbing bumper, airplane wing, concrete roadbed, artificial heart valve, or component in an integrated circuit. In general, the study of materials performance strongly overlaps with design. Although synthesis and processing are often thought of in serial fashion (i.e., the equipment or product designer specifies needed properties, and materials personnel then choose or develop materials that have those properties), it is becoming increasingly common to consider the synthesis and processing and the behavior of materials as an integral part of the device or equipment design process. In all real applications, many materials properties play roles in the design of a system. For example, in a relatively simple system such as an automobile hood, relevant properties include not only density, corrosion resistance, strength, stiffness, and forming and welding parameters, but also electrical conductivity (because of the possibility of radio frequency interference between engine devices and the antenna) and magnetic properties (which come into play in the separation of iron for recycling after use). More complex combinations of materials and properties interact in subtle ways in most machines, devices, and components, from engine components to electronic packaging elements.

Effective research programs bearing on performance require broad interactions of materials science and engineering researchers with participants from manufacturing and other engineering disciplines. Except in some industrial and federal laboratories, such interactions do not occur effectively at present. Indeed, performance-oriented research is not adequately pursued today in academic materials science departments in the United States—a shortcoming, given its importance to materials science and engineering. This lack of attention by academia seems to revolve around the perception that performance-oriented research is "too macroscopic" or "not fundamental enough." But performance-oriented research involves many intellectually challenging problems, ranging from understanding the microstructural response of interfaces between complex solids to predicting lifetimes for structural materials subject to stress or corrosion. Progress in these areas depends on tools and perspectives drawn from all of traditional materials science and engineering plus related aspects of chemistry, physics, mathematics, and engineering.

Performance is similar to synthesis and processing in that it has been underemphasized in the United States, particularly in federally funded research in areas relevant to commercial issues. Additional efforts to evaluate and predict the performance of materials in the context of their use have the potential to contribute substantially to materials problems of economic importance. Design of materials for improved performance relies on improved experimental techniques and improved theoretical understanding of such performance-related properties of materials as susceptibility to mechanical fatigue, an example of which is shown in Figure 4.2, which illustrates cracking. Performance research will be carried out best at facilities where creative approaches to synthesis and processing, characterization, design, and performance are found and where cooperative mechanisms exist to draw these elements of materials science and engineering together.

The interaction between materials synthesis, materials performance, and component or equipment design is becoming increasingly sophisticated and complex. It is further complicated by the fact that current and future products increasingly call for intimate combinations of novel materials. Optimum trade-offs have to be found in design versus performance. To find them by older empirical methods would generally be too time-consuming and costly. Instead, increasing use must be made of analysis and modeling, which often call for intensive computing capabilities. Likewise, there are growing demands for improved instrumentation, particularly for monitoring or deducing the performance of materials in their design environment.

STRUCTURE AND COMPOSITION

A given material contains a hierarchy of structural levels, from the atomic and electronic to the macrostructural level. At all of these structural levels, chemical composition and distribution may vary spatially. These structures and compositions are the result of the synthesis and processing that have been applied to make a given material. In turn, the nearly infinite variety of possible structures gives rise to the similarly complex arrays of properties exhibited by materials. Because of the fundamental role of structure and composition, understanding them at all levels is an essential aspect of materials science and engineering.

Critical to the rapid advances in materials science and engineering over recent decades has been the development of continually more powerful tools for probing structure and composition. Fifty years ago, the major tools used were the light microscope, x-ray diffraction, and infrared and ultraviolet spectroscopy. Today, researchers have access to an enormous range of new instruments and technologies. The scanning tunneling microscope enables

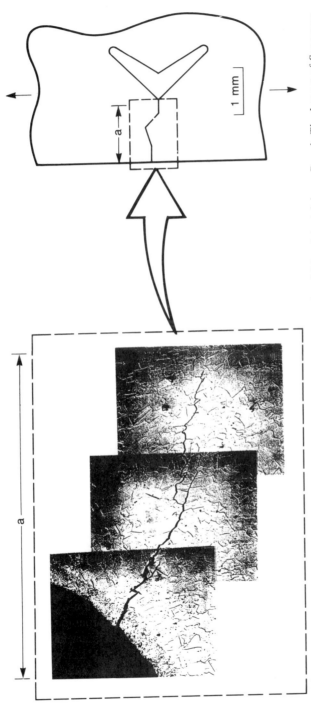

FIGURE 4.2 Left: Micrograph of a typical surface flaw in stainless steel. Right: Schematic of the plastic zone at the tip of a flaw. The nature and extent of the plastic zone determine a material's fracture resistance and consequently its performance in service. (Reprinted from National Materials Advisory Board, *The Impact of Supercomputing Capabilities on U.S. Materials Science and Technology*, National Academy Press, Washington, D.C., 1987.)

determination of atomic arrangement and electronic structure at and near the surface of materials. Solid-state nuclear magnetic resonance allows determination of chemical makeup in complex polymer systems. Electron microscopy can show atomic arrangements and chemical compositions at near-atomic resolution; at lower magnifications, it allows the determination of maps of chemical inhomogeneity on larger scales. A host of spectroscopies enable the chemical characterization of surfaces. High-intensity neutron beams from reactors and photon beams from synchrotron sources have made possible a vast array of techniques for chemical and structural characterization.

This new instrumentation has increased understanding but brings with it major concerns for the field of materials science and engineering. The high cost of characterization has become an issue demanding care in the balance of allocations by funding agencies and has, in many instances, become a limiting factor in the progress of research. Furthermore, the availability of large characterization facilities in only a few geographic locations has led to changes in the way research is carried out for many materials scientists and engineers. Rather than do these experiments in their own laboratories, they may now travel thousands of miles to do their research at a major national facility in the midst of the exciting intellectual ferment present at such facilities.

Concurrent with the development of techniques for characterizing the structure and composition of materials has been the development of analytical and modeling techniques to explain the origins of these observations, for example, quantum calculations to describe electronic structure and crystal structure stability; equilibrium and nonequilibrium thermodynamics to describe multiphase materials; and hydrodynamics and instability analysis to explain the development of microstructures in crystalizing metals and polymers.

Historically, the development of materials has involved many key discoveries made at the macroscopic level (such as continuum behavior or mechanical properties). In recent decades, there has also been increased emphasis on the microscopic or atomic level, both in research and in education. New generations of electrical engineers, ceramists, metallurgists, polymer chemists, and condensed-matter physicists, who have been trained in the basic interactions of atoms and molecules, understand the fundamental concepts underlying and unifying previously disparate classes of materials. Opportunities to increase fundamental understanding in this area continue to occur. Three years ago, for example, the apparently well-defined science of geometric crystallography was jolted by the discovery of icosahedral symmetry in solids. These so-called quasi-crystals possess orientational order without translational periodicity. Study of these quasi-crystals promises to lead to a deeper understanding of the conditions leading to different atomic arrangements in solids. Figure 4.3, a picture of icosahedral Al_6Li_3Cu, shows the triacontahedral faceting that occurs upon slow quench of the phase. Figure

FIGURE 4.3 Icosahedral Al_6Li_3Cu. (Reprinted, by permission, from P. A. Heiney, P. A. Bancel, P. M. Horn, J. L. Jordan, S. LaPlaca, J. Angilello, and F. W. Gayle, 1987, Disorder in Al-Li-Cu and Al-Mn-Si Icosahedral Alloys, Science 238:660–663. Copyright © by The American Association for the Advancement of Science.)

4.4 shows three slices of a three-dimensional icosahedral quasi-crystal; pieces include a rhombic triacontahedron.

A new emphasis in materials science and engineering centers on the nanometer size regime, which is intermediate between the well-studied macroscopic and atomic levels. This regime is pivotal in understanding the magnetic, electronic, and optical properties of materials. Knowledge of structural and compositional features in the nanometer size range is also important for interfaces between dissimilar materials such as those that occur in composite structural materials or in the complex multimaterial devices that make up integrated electronic circuitry. Studies of synthesis and processing have focused increasingly on the nanometer size regime, as is demonstrated by development of block copolymers, ultrafine ceramic powders, metallic microstructures, and superlattice electronic devices. The increased understanding of materials at this level has shifted the fundamentals of process control

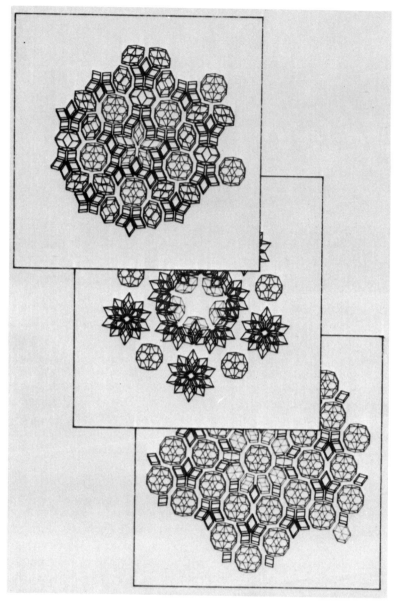

FIGURE 4.4   Three slices of a three-dimensional icosahedral quasi-crystal. (Reprinted, by permission, from Paul Steinhardt. Copyright © 1989 by Paul Steinhardt.)

away from purely macroscopic phenomena. Moreover, numerous new synthetic techniques have made it possible to fabricate nanometer-scale structures that exhibit new physical phenomena, which in turn can form the basis for new technologies. Increasingly, the properties and performance of materials are determined by the nanostructure of the materials, and society is using more of these materials each year. Development of such materials represents a scientific frontier with technological importance to many industries.

SYNTHESIS AND PROCESSING

Synthesis and *processing* are terms that refer to the building of new arrangements of atoms, molecules, and molecular aggregates; the control of structure at all levels from the atomic to the macroscopic; and the development of processes to produce materials and components effectively and competitively. *Synthesis* is often used alone to refer to the physical and chemical means by which atoms and molecules are assembled. *Processing* may be used in a similar way, for example, in the phrase *electronic materials processing*. Processing may also imply changes on a larger scale, including materials manufacturing. It is often applied to such macroscopic manipulations as ingot solidification, mechanical modification, sintering, and joining. These macroscopic manipulations, of course, also cause important structural changes at the levels of atoms and grains.

In materials science and engineering, the distinctions between synthesis and processing have become increasingly blurred in recent years. The fabrication of artificially structured materials, which involves synthesis of materials on the atomic scale, is typically referred to as processing. The preparation of ceramics, which in the past generally involved sintering of mixtures of mineral-derived oxides, now involves considerable synthetic chemistry in some instances. Broadly, it may be stated that synthesis and processing form a continuous range of activities in which assemblages of atoms, molecules, and molecular aggregates are transformed into useful products.

Synthesis and processing research is evolving to the point that, in some cases, new materials can be tailored, atom by atom, to achieve a desired set of properties or to obtain new and sometimes unexpected phenomena. Synthesis and processing encompass a comprehensive array of techniques and technologies as diverse as rolling of sheet steel, pressing and sintering of ceramic powders, ion implantation of silicon, creation of artificially structured materials, ladle-refining of steel, sol-gel production of fine ceramic powders, pouring of polymer-modified concrete, shaping by machining or chip processes, thermomechanical processing of alloys, preparation of polymers by chemical reactions, coating of turbine blades for corrosion resistance, zone refining of silicon, growth of gallium arsenide crystals, and laying-up of composite materials. Some of these technologies are quite new and may, in time, lead to major technological advances and industrial growth. Others are well embedded in established industries but require continual improvement if the U.S. industries that rely on them are to remain competitive with foreign industries.

Synthesis and processing are key to production of high-quality, low-cost products throughout a broad spectrum of manufacturing. They are central to translation of new research and new designs into useful devices, systems, and products. They are essential to efficient introduction of advanced materials or materials combinations into the marketplace. A case in point are the ceramic components used in catalytic converters. A key element in many catalytic converters is the monolithic substrate (Figure 4.5). The substrate consists of several ceramic layers, the final one added to assure a high surface area. The catalyst, composed of precious metals such as platinum, palladium, and rhodium, is applied as a thin coat about 30 to 50 μm thick, and the assembly is incorporated as the active component into the catalytic converter.

Synthesis and processing also represent a large area of basic research in materials science and engineering. From this basic research come wholly new materials, for example, new conductive polymers, new compositions

FIGURE 4.5 Monolithic substrate configurations developed as active components in catalytic converters. The circular and oval-shaped versions are currently preferred shapes. (Reprinted, by permission, from Ford Motor Company. Copyright © 1989 by Ford Motor Company.)

of ceramic superconductors, dislocation-free single crystals, and artificially structured materials. Another important thrust of basic processing research is to develop a fundamental understanding of kinetic phenomena involved in materials processing, to serve as the foundation for changes and improvements in processing. Examples of such phenomena are rheological behavior in die filling, atomistic mechanisms of crystal growth, atomistic mechanisms of removal of materials in machining, and mass transport mechanisms in consolidation processes.

The United States suffers from a serious weakness in synthesis and processing with respect to new materials, manufacturing technology, and education in materials science and engineering. In many areas, the synthesis and processing of materials have been emphasized less in government, industrial, and university laboratories in the United States than in laboratories of other countries. Not only does this sharply limit the techniques that can be brought to bear on problems in this area, but it also curbs opportunities for unexpected discoveries.

An important but often overlooked aspect of synthesis and processing advances is the continued development of new machinery and equipment for synthesis and processing. As discussed below in the section "Instrumentation," research devoted to equipment development receives only limited support in the United States, with a commensurate loss of equipment markets to foreign competition. Notable examples in process technology include the markets for machine tools and semiconductor processing equipment. To ameliorate the flow of resources for manufacturing equipment to foreign markets, the United States needs to accelerate research on synthesis and processing equipment and to strengthen the manufacturing industry for this equipment. A strong machine and equipment component is essential to improving the synthesis and processing component of the technology base for any industry.

It is also important to improve the scientific foundations underlying U.S. manufacturing processes. Processing efficiency not only is essential to U.S. industrial competitiveness, but also poses many intellectual problems. Basic research directed at increasing the understanding of crystal growth, vapor deposition, sintering, phase transformations, rheology, and other generic processes key to manufacturing could have profound effects on national productivity. This research will be particularly valuable when it can be extended to the development of real-time process models that can be used in process control. In recent years, robust nondestructive sensors have been developed to measure materials properties during processing. For example, sensors are now available to check the thickness, grain size, and texture of thin-sheet-rolled metals as they are being processed. Figure 4.6 depicts a sensor that measures various properties of an aluminum rod during extrusion processing. The introduction of such sensors earlier in the process stream

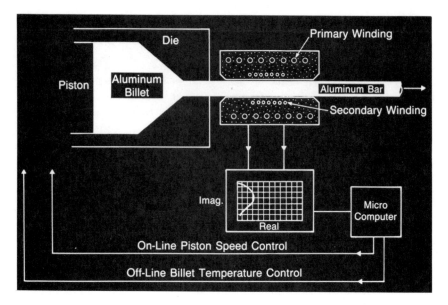

FIGURE 4.6 Schematic representation of a prototype eddy current sensor for measuring the diameter, electrical conductivity, and temperature of an aluminum rod during extrusion processing. The control system uses sensor-acquired temperature measurements in a feedback loop, performing off-line control of the initial temperature of the billets and on-line control of the speed of extrusion (itself a heat-generating process). This system will result in improved product quality and reduction of rejected output through in-process temperature and measurement control. (Courtesy National Institute of Standards and Technology.)

enables improved control and/or rejection of bad products at minimum value-added process steps. When coupled with process modeling and elements of artificial intelligence, these sensors promise to usher in a new era of intelligent processing of materials. Developments in this field will require the combined efforts of experts in sensor development, materials modeling, the relationships between process variables and product structure, and artificial intelligence (with a strong emphasis on expert systems). To maximize the impact of such an effort, collaboration between industrial, university, and government laboratories must be achieved at the outset. An expanded effort by funding agencies and national laboratories directed to the intelligent processing of materials could have an immediate impact on the quality of processing of conventional materials and could hasten the introduction of advanced materials. Such an effort would also have the benefit of focusing academic attention on processing and of stimulating the interest of a new generation of highly qualified engineering and science students in this vital area.

In the United States, a long-standing tendency to view synthesis and processing as a service function that is not an elite activity of science or engineering has often led to a weak linkage between synthesis and processing and the other elements of materials science and engineering. Even though this viewpoint is generally understood to be invalid in today's world, it persists in the structure of scientific institutions—particularly of universities—and in many industries, where production is not viewed as a route to senior positions. Furthermore, because academia has often given comparatively little attention to synthesis and processing, there has been a shortage of qualified scientists and engineers in these areas (this issue is discussed further in Chapter 5). There is no more practical argument for the necessity of a unified view of materials science and engineering than this relative neglect of synthesis and processing. This lack of support exists not only in the federal government and universities but also in many industries, and it spans the complete range of materials-related activities, from science to engineering and from the creation of new materials to the manufacture of products using materials.

Because of the crucial role the committee sees for synthesis and processing, the remainder of this chapter provides examples of areas for pioneering research in synthesis and processing, most of which extend across several of the materials classes.

Artificially Structured Materials

The development of new materials and of materials systems structured on an atomic scale is a recent phenomenon that will have many important applications. For example, the processes used for producing artificially structured materials make it possible to combine optically active materials with electronic circuitry in ways that should lead to qualitatively new kinds of optoelectronic devices. Artificially structured materials can be produced by a variety of techniques, including molecular beam epitaxy (MBE), liquid-phase epitaxy (LPE), chemical vapor deposition (CVD), vacuum evaporation, sputter deposition, ion beam deposition, solid-phase epitaxy, chemical beam epitaxy (CBE), metallo-organic molecular beam epitaxy (MOMBE), and low-pressure chemical vapor deposition (LPCVD).

Techniques for growing thin films epitaxially, such as MBE, have been used to produce artificially structured materials with levels of purity and structural perfection that seemed impossible only a few years ago. Layered semiconductor systems with layer thicknesses of atomic dimensions and with atomically smooth interfaces are now grown. The gallium-arsenide/gallium-aluminum-arsenide (GaAs-GaAlAs) system has received the most attention to date, with structures of widely varying electronic properties produced by control of composition. Carrier mobilities in excess of 10^6 cm^2/V-s have been achieved in systems with layer thicknesses of the order of 10^{-6} cm.

An artificially structured material generally can be expected to exhibit novel and useful properties when the length scale of the structure is comparable to the characteristic length scale of the physical phenomenon of interest. Examples of interesting microscopic length scales include the de Broglie wavelengths of electrons, the wavelengths of phonons, the mean free paths of excitations, the range of correlations in disordered structures, characteristic diffusion distances, and the like. These distances can vary from a few atomic spacings to microns. The least explored area, and the one with the greatest potential interest for processing technology, lies between the atomic and macroscopic sizes.

To date, most of the interest in the field of artificially structured materials has been focused on semiconductors, but there are many opportunities for new and useful combinations involving metals, insulators, and even polymers. The processing technologies now available are capable of producing both equilibrium and novel nonequilibrium phases, including amorphous structures and extended solid solutions. The range of possibilities in this area is truly remarkable.

Ultrapure Materials

New synthetic methods and new procedures for handling materials during synthesis are now yielding substances of unprecedented purity and performance. The synthesis of very pure substances is becoming increasingly important both in microelectronics and in the development of new structural materials. The need for extremely pure silicon in microelectronic devices is well known. More recently, very pure and atomically perfect III-V semiconductor crystals are being used in advanced electrooptical devices. In the area of structural materials, it is becoming clear that oxygen and carbon impurities can limit the strength of fibers used in composites. Crack initiation and growth in solids can be attributed to impurities and defects.

The success of molecular precursors in solid-state synthesis often depends on the use of ultrapure molecular materials. For example, a promising new method for producing ceramic fibers starts with synthesis of a preceramic polymeric material that can be processed into a fiber and then pyrolyzed to form the ceramic. The purity of the preceramic polymer determines the strength of the ceramic fiber. In another example, very pure organometallic precursors can be used in synthesizing complicated ternary and quaternary III-V compounds for use in the preparation of device-quality materials. Lasers with unusually low threshold currents have been produced in this way. Finally, molecular precursors to metal lines and thin films may be useful in device fabrication.

Organic nonlinear optical materials provide another illustration of the opportunities for research in the synthesis of ultrapure substances. It is becoming widely appreciated that the nonlinear optical properties of organic and organometallic molecules can be superior to those of inorganic solids. Preparation of new organic and organometallic molecular solids for use in nonlinear optics represents a special opportunity for academic chemists, because the synthetic techniques are relatively commonplace in major chemistry departments in the United States. What makes this a new opportunity is the need to prepare organic and organometallic solids of a purity and optical quality seldom demanded in typical chemical applications. New strategies are needed for designing and preparing organic and organometallic solids with good optical properties.

One of the most critical and pervasive needs for ultrapure materials is in fundamental physical studies of structure-property relationships. For example, very pure samples of polymers with narrow distributions of molecular weights are needed to test modern theories of the behavior of polymers in dense fluids. Thus, for both fundamental and applied purposes, laboratories dedicated to the preparation of ultrapure materials must have high priority in programs aimed at upgrading U.S. capabilities in materials science and engineering.

In metals, the importance of purity with respect to ductility, resistance to corrosion, strength, and other properties is increasingly evident. Toughness at low temperatures (e.g., in steel for pipelines in cold climates) is obtained through the use of high-purity carbon steels with very low sulfur and phosphorus levels. Vacuum induction melting, vacuum arc remelting, and electron beam melting are used to obtain structural metals (especially superalloys and titanium) of extremely low impurity contents. Reduction of iron impurities significantly below usual commercial specifications results in dramatic improvement in the corrosion resistance of magnesium alloys. Purification of copper through zone melting imparts to the metal the ductility necessary to draw it economically to the extremely fine diameters needed for interconnects in very large scale integrated circuits.

New Structures

The systematic search for new and potentially useful structures, based on theory, empiricism, or usually a combination of both, is an essential activity of materials science and engineering. It is the domain of synthetic chemists, who work to develop new polymeric molecules, ceramic superconductors, and fast-ion conductors. It is also the domain of solid-state scientists, processors, and others, who find and develop, for example, new metastable structures, structures that provide new strengthening mechanisms, and new superhard materials.

History shows that the search for radically new structures is an important goal for materials science and engineering, even when no immediate application is in sight. Fifty years ago, scientists in England and Germany were developing techniques to obtain fully columnar microstructures in castings; today, these same techniques are used in production of the most advanced turbine blade materials. The quasi-crystalline structures, the nano-crystalline structures, and the new metastable structures being produced today may find equally important uses in the decades ahead.

Solidification

Most engineering materials pass through the molten state at some stage during their processing, and the transformation from liquid to solid is an area ripe for fundamental research and technological development.

Nucleation and growth processes have been important areas of research for more than 30 years. Today, important research topics on nucleation deal with how to avoid it to achieve high undercoolings and, hence, nonequilibrium structures in materials, and with how to promote it to achieve fine grain sizes. Topics in growth deal with interfacial phenomena, dendritic growth mechanisms, nonequilibrium processes, and formation of heterogeneities such as lattice defects, segregation, porosity, and inclusions.

In recent years, great advances have been made in growing single crystals; a notable example is provided by the semiconductor industry, in which, during the last 20 years, the size of silicon single crystals grown from the melt has increased from 1 in. (2.5 cm) in diameter to more than 6 in. (15 cm) in diameter (with further increases expected in the future), while the dislocation content of these crystals has dropped from 100 to $1000/cm^2$ to practically zero.

Solidification processing can be configured to achieve not only the desired bulk material properties but also components in almost the desired form (a process known as near-net-shape forming). Newer foundry techniques often exploit solidification processing to achieve reduced costs. But the process is also used to achieve special structures with unique properties. One example is the directionally solidified turbine blades used on high-performance jet engines (see Figure 2.1); another is composites made by semisolid forming. Casting of steel strip is a continuous near-net-shape casting process that is attracting worldwide attention. It is used today for metals with lower melting points, and it appears to be only a matter of time until it is perfected for steel.

In conventional casting processes, cooling rates are on the order of 1 K/s, and they are much lower in some cases. In rapid solidification processes, cooling rates of from 10^2 to 10^8 K/s are obtained; at the higher ends of the scale, crystallization may be wholly prevented, even in metals and low-viscosity ceramics. A process for making rapidly solidified powder is illustrated in Plate 3. A more detailed discussion of the promises of rapid solidification processing is given in Appendix B.

Rapid solidification technology has led to amorphous materials with new and useful combinations of magnetic properties. Their unique soft magnetic properties will lead to applications in electronics, power distribution, motors, and sensors. New permanent magnets produced by rapid solidification will be useful in building compact, powerful motors. Rapid solidification has also led to new fine-grained and homogeneous crystalline materials with improved properties and performance. The materials that have responded well to this processing technology include high-strength aluminum and magnesium alloys, tool steels of high toughness, nickel-based superalloys, and oxide abrasive materials. Many thousands of tons of rapidly solidified alumina zirconia abrasives are now produced and sold each year.

Rapid solidification recently played a key role in the remarkable discovery of the so-called quasi-crystalline phases. These phases were first produced accidentally during rapid solidification of aluminum-manganese alloys. The scientific interest in these phases arises from the fact that they display long-range order—they are not amorphous or glassy—but the symmetry of the order is not consistent with the heretofore accepted rules defining the allowable symmetries of crystals. The discovery of quasi-crystals has led to an ongoing reexamination of the basic principles of crystallography, a science that now will have to be reformulated in a more general framework. It is not known, at present, whether these new phases will have interesting and useful properties, but this entirely new phenomenon clearly calls for intense investigation. It is notable that a study of structure and properties made possible by rapid solidification processing has led to a major discovery in crystallography.

Vapor Deposition and Surface Processing

Vapor-solid processing is becoming an increasingly important tool for achieving ultrafine structures, epitaxial layers, surface coatings, and bulk forms in single shapes. The list of processes used is very long and includes physical vapor deposition, CVD, plasma-assisted vapor deposition, metallo-organic chemical vapor deposition, MBE, and ion beam deposition. Vapor deposition processes are used extensively in the electronic materials industry to build chip structures. They also have wide applications in other areas.

Chemical and physical vapor deposition processes have long been used to coat high-temperature materials, notably turbine blades. Plastic parts are sometimes coated with a metal by vapor deposition so that they appear to be metallic. The backs of gemstones are sometimes coated with a metal to increase their luster. Vapor processes are involved in metal reduction or purification of many metals. Solid shapes (e.g., of "pyrolytic" carbon) are formed by "vapor forming."

Many broad research opportunities evident in this field involve nucleation, growth, and materials transport. These processes can be carried out far from equilibrium, providing the opportunity to obtain new and nonequilibrium structures and compositions. Co-deposition processes can produce unique and layered structures. Most vapor-coating processes involve some degree of intermixing of the coating with the substrate; this intermixing is enhanced in processes such as ion beam deposition that provide new approaches to fabricating new structures.

Solid-State Forming Processes

Solid-state forming processes include those that involve extensive materials flow (e.g., injection molding, rolling, forging, and calendaring) and those that involve cutting and grinding. Important new processes that continue to be developed in these areas can influence productivity in important ways and can also result in production of materials with new structures and new properties. Figure 4.7 illustrates technologies for studying strain in deformed metal sheets.

Continuous rolling and annealing of steel sheet has been made possible in recent years by advances in equipment and especially in instrumentation and computer technology. The result is less expensive sheet of greatly improved quality. Injection molding, previously the domain of polymers, is being used increasingly for ceramics and metals. Forming of semisolid (semimolten) metals is a new process, now reaching commercialization, that is a hybrid between forging and casting. In addition to process innovation and process control, process modeling presents important opportunities for research in this field.

Joining, Consolidation, and Materials Removal

Joining processes range from welding of structural steel plate to soldering of the nearly countless interconnects between electronic chips. Both extremes provide important areas of research in process fundamentals and process innovation. Figure 4.8, which shows an example of process fundamentals, illustrates the effects of substrate crystallography and beam direction on the microstructure of an electron beam weld.

Consolidation processes include processing and manufacture of composite materials by a variety of techniques, especially pressing and sintering. Im-

FIGURE 4.7   Metal coupons with dot patterns such as these are used to measure the strain in a deformed metal sheet up to (top) and including (bottom) ductile tearing. (Reprinted from National Materials Advisory Board, *The Impact of Supercomputing Capabilities on U.S. Materials Science and Technology*, National Academy Press, Washington, D.C., 1987.)

portant new consolidation processes involve hot pressing, hot forging (as applied to ceramics), and hot isostatic pressing.

Materials removal involves a wide range of processes, including cutting, grinding, machining, etching, and ion bombardment; it ranges in scale from the scarfing of large ingots to the manufacture of semiconductor chips. Lasers, advanced ceramic materials, and new synthetic diamond materials now provide a new dimension for cutting and grinding.

Important research topics, many with far-reaching economic implications for a range of materials, exist in all these areas.

Electrolytic Processing

Electrolytic processing is an important segment of the broader, $28 billion electrochemical industry. The processing segment includes metal production, plating, semiconductor processing, and chemical production. In recent years, new materials, new processes, and process modeling have had a dramatic influence on the course of the industry, and they promise to further alter its character in the years to come.

An important topic for research in electroprocessing is the electrosynthesis of advanced materials. Co-deposition is one method. For example, the ternary

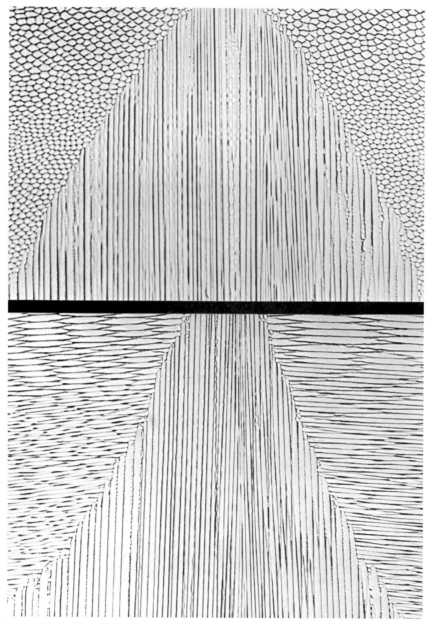

FIGURE 4.8   Electron beam welds made along the (110) (top) or (100) (bottom) directions of a single crystal Fe-^{15}Ni-^{15}Cr alloy, revealing the influence of beam direction and crystallography on the solidification substructure (original magnification: ×500, top; ×400, bottom). (Courtesy Oak Ridge National Laboratory, operated by Martin Marietta Energy Systems, Inc., for the U.S. Department of Energy.)

magnetic alloy neodymium-iron-boron has been produced by co-deposition from a molten salt. A wide range of compounds—from electronic and photonic materials such as gallium arsenide and cadmium telluride to wear-resistant materials such as tungsten carbide to electron emitters such as lanthanum hexaboride—have been deposited. Opportunities exist to develop new nonaqueous chemistries to extend the window of electrochemical potential and thereby give greater access to highly reactive materials.

Electrolysis is a nonequilibrium process that has the capability of generating nonequilibrium structures such as coatings, epitaxial layers, or powders. Compositionally modulated structures such as that illustrated in Figure 4.9 are readily produced electronically.

Electrodeposition is carried out at or near room temperature in aqueous and organic electrolytes, at elevated temperatures in molten salts, and at low temperatures in cryogenic liquids. Cryogenic electroprocessing offers a new window of opportunity in a temperature region in which kinetic processes occur at rates quite different from those that occur at room temperatures or elevated temperatures, providing new opportunities for achieving metastable structures.

COMMON THEMES

Two important themes cut across all four of the elements of materials science and engineering. The first is the importance of instrumentation in performing and controlling synthesis and processing; characterizing structure, composition, and properties; and analyzing performance (see Appendix D). The second is the increasing importance of analysis and modeling (see Appendix E), which, together with increased computational abilities and judicious experimentation, are helping to make materials science and engineering increasingly quantitative.

Instrumentation

The instrumentation used in materials science and engineering has become increasingly sophisticated and expensive. The cost of analytical instruments such as electron microscopes ranges from $50,000 to $850,000, and the cost of dedicated process equipment, such as MBE and ion implantation equipment, is in the range of $1 million to $2 million. Furthermore, these costs are rising rapidly, at well above the rate of inflation. With the increasing use of powerful computers to simulate the synthesis and processing, structure and composition, properties, and performance of materials, instrumentation has become important for theoretical as well as experimental materials research. Cost-effectiveness rules out having such facilities at every institution, making new sharing mechanisms a necessity.

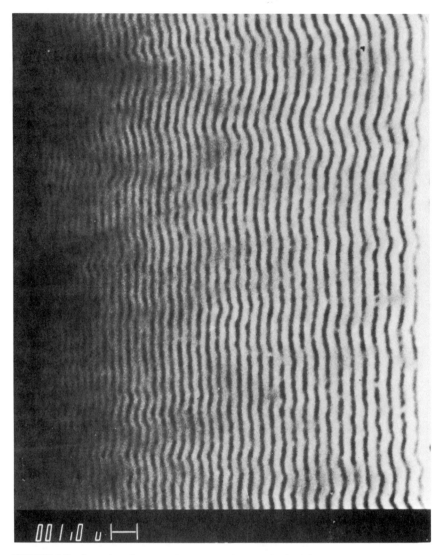

FIGURE 4.9   Scanning electron microscope micrograph of an electrochemically produced copper-nickel compositionally modulated alloy grown on a (111) copper single crystal with the layer thickness continuously graded from 15 nm to 1500 nm. This gradation in layer thickness results in gradations in those properties—such as flow stress, hardness, corrosion resistance, magnetic properties, and wear resistance—that depend on the layer thickness. (Courtesy National Institute of Standards and Technology.)

Low levels of research support during the past decade and a half, as described in Chapter 6, have made it impossible for universities to modernize their facilities or even to maintain and use existing facilities effectively. As a result, much of the equipment used for basic research in materials science and engineering is quite old and not always fully functional. According to the 1986 report of the White House Science Council's Panel on the Health of U.S. Colleges and Universities, more than 45 percent of the instrument systems for materials science at universities are more than 10 years old, about 20 percent are between 6 and 10 years old, and about 35 percent are from 1 to 5 years old. In fact, the instrument systems for materials science were older than the systems for any other scientific area sampled. This agedness of materials research equipment in universities is a major problem.

Very recently, the National Science Foundation (NSF), the Department of Energy (DOE), and the Department of Defense (DOD) have undertaken special initiatives to upgrade instrumentation in materials research. In FY1987, the NSF spent almost 17 percent of its budget for research facilities, and its division of materials research devoted about 22 percent of its budget—or about $22 million—to these purposes. Other agencies are also devoting an increasing fraction of their resources to instrumentation. Yet these amounts are dwarfed by the magnitude of the need. According to the aforementioned White House panel study, the materials research community in universities requires $100 million to $400 million in new instrumentation to replace aging equipment and to address new opportunities.

In addition to the need for instrumentation to support ongoing materials research at universities, there is a pressing need in the United States for the development of new instrumentation. Increasingly, the development of new instruments is taking place in foreign countries. American laboratories now depend primarily on foreign suppliers for various types of essential equipment, ranging from apparatus used for growing crystals to electron microscopes and superconducting magnets used in a variety of spectroscopic and low-temperature measurements. In these areas of research, more instrument companies exist in other countries than in the United States, and these companies work more closely with government laboratories and universities. In these same areas, national laboratories abroad also play a much more important role in instrument development than they do in the United States.

A particular example from surface science is relevant. Twenty years ago, the United States was the dominant player in the development of surface science instrumentation, as demonstrated by development of the Auger spectrometer and the low-energy electron diffraction display system. Universities, industrial laboratories, and instrument companies were all actively involved in the development of such instruments. Since then, the United States has lost almost completely the dominance it once enjoyed. Innovative designs or unique applications of instruments are still being pursued in the United

States, but these developments are not being converted into commercial instruments. The commercial development of instrumentation is occurring increasingly in foreign laboratories able to sustain complicated and expensive instrument development programs.

Another important example involves processing equipment. Research on processing equipment in many technologies related to materials science and engineering has subsided in the United States, with a commensurate loss of equipment markets to foreign competition. Notable examples include the markets for machine tools and semiconductor processing equipment. Yet a leading position in materials processing technology requires leadership capability in the machinery and equipment sector and a close collaboration with materials processing. Figure 4.10, a scanning electron micrograph of polystyrene, illustrates the use of advanced instrumentation to explore materials morphology and growth processes. Figure 4.11, which shows fiber pullout and bridging across a crack in a fiber-reinforced resin, illustrates the importance of microscopic analysis in performance research.

FIGURE 4.10 Scanning electron microscope micrograph of rosettelike crystalline aggregates of isotactic polystyrene grown from dilute polymer solutions. The splaying-layered morphology of these objects has a direct bearing on aspects of the growth processes that govern the morphological evolution of spherulites of this polymer grown from a molten state. (Courtesy National Institute of Standards and Technology.)

FIGURE 4.11   Fiber pullout and bridging across a crack in a fiber-reinforced resin. (Courtesy E. I. du Pont de Nemours & Co., Inc.)

Federal funding agencies have not made instrument development a prominent component of their programs. The Division of Materials Research at the NSF spends only about 1 percent—about $1 million per year—for the development of new instrumentation. The DOE supports instrument development only on a limited, project-specific scale, and the DOD has no documented case of instrument development in materials science having been funded through its University Research Initiative program. This neglect of instrument development ignores the research leverage resulting from early access to novel equipment.

Several consequences are likely if instrument development is not pursued in American universities. First, the lag between demonstration of a new instrument and its transfer into American research or industrial laboratories will increase. In surface science, there is often a 5- to 10-year lag between an instrument's demonstration in a foreign laboratory and its availability in the United States from a commercial instrument company. Second, fewer students will be trained in instrument development. Third, foreign companies will command an increasing portion of the commercial instrument business, especially for small, specialized equipment.

Many scientists and administrators contend that instrument development is not compatible with the operation of American universities. The period for promotion to tenure might be too short for young scientists embarking on a difficult instrument development program, and a project might require more time than a graduate student's education. But these problems can be overcome, as demonstrated by laboratories in Europe and Japan and by different departments (such as high-energy physics, astrophysics, or biology) within the American university system.

The potential for development of new instrumentation is particularly great at present. For instance, materials processing equipment could combine computer control of such technologies as MBE, plasma deposition, and particle beam lithography. Synergistic combinations of emerging technologies could give birth to new and exciting instruments.

Analysis and Modeling

Recent advances in theoretical understanding and computational ability are changing the nature of materials science and engineering. Two complementary forces are driving these changes. The first is the unprecedented speed, capacity, and accessibility of computers. Problems in mathematics, data analysis, and communication that seemed untouchable just a few years ago can now be solved quickly and reliably with modern computational systems. The second is the growing complexity of materials research. The latter change has occurred in large part because instruments are now available to make highly detailed and quantitative measurements and because the computational ability exists to deal with the resulting wealth of data. Underlying all of these developments are advances in the theoretical understanding of materials properties and the mathematical ability to devise accurate numerical simulations.

Analysis and modeling find application in each of the four elements of materials science and engineering. For instance, in performance research, advances in analysis and modeling have made it possible to develop quantitative methods for solving some of the major problems in the field. In particular, accurate codes may become available to enable reliable predictions of the lifetimes of structural materials in service. Such observations also apply to the application of analysis and modeling in synthesis and processing and in characterization of the structures, compositions, and properties of materials. An example of computer analysis of copolymer systems is shown in Plate 4.

Analysis and modeling in materials research traditionally have been divided into roughly three different areas of activity. The most fundamental models—those used primarily by condensed-matter physicists and quantum chemists—deal with microscopic length scales, where the atomic structure of materials plays an explicit role. Much of the most sophisticated analysis is carried out at intermediate length scales, where continuum models are appropriate. Finally, modeling of complex industrial processes aids in improving process reliability and control, and in reducing an important aspect of performance— the initial cost of the material or components. There is work done at macroscopic length scales in which the bulk properties of materials are used as inputs to models of manufacturing processes and performance. Historically, research in each of these three areas has been carried out by separate communities of scientists. However, modern developments are tending to blur the distinctions between these research areas.

The need for advanced analysis and modeling provides an especially clear argument in favor of unified support for materials research. For example, in the past it has sometimes been difficult for managers of engineering programs to support accurate experimental or theoretical investigations of simple model systems; both the models and the precision appeared to be irrelevant for technological purposes. At the same time, many such projects have seemed scientifically uninteresting in the absence of a technological motivation. It is now apparent, however, that the simulations needed for advanced applications may make no sense without scientific input. Carefully controlled measurements, critically evaluated data, and calculations are essential to test the basic assumptions being built into simulations of complex situations. Moreover, the problems often turn out to be unexpectedly challenging from a scientific point of view.

In the future, science-based numerical simulations in combination with new methods for storing, retrieving, and analyzing information may make it possible to optimize not only the properties of specific substances but also entire processes for turning raw materials into useful objects. Materials considerations are important throughout the life cycles of most products—from design through manufacturing to support and maintenance and, finally, to disposal or recycling. Significant improvements in quality, reliability, and economy might be realized at all stages of this cycle if quantitative models of processing and performance could be applied throughout this process.

FINDINGS

Materials science and engineering deals with the synthesis and processing, structure and composition, properties, and performance of materials and with the interrelationships among these elements of the field. It encompasses all materials classes. To a degree unique among technical fields, it gains great strength from drawing on the full spectrum of science and engineering.

• In evaluating the status of materials science and engineering, it is important to recognize not only that each of the elements of the field must be strong, but also that the elements of the field are increasingly interdependent. Only by viewing the field as a coherent whole can the interrelationships among its elements be distinguished and strengthened.

- Synthesis and processing is an element of materials science and engineering offering great promise as new techniques for the preparation of materials enable nearly atom-by-atom synthetic flexibility. It is an element of special importance also, because it bears directly on questions of industrial productivity and international competitiveness. However, the United States suffers from a serious weakness in synthesis and processing, particularly in that aspect required to translate scientific promise into commercial success: process technology.

- Structure and composition can now be characterized with unprecedented accuracy and resolution. The challenge is to give practitioners of materials science and engineering broad access to the increasingly expensive equipment—including major national facilities—required to perform this characterization.

- Unprecedented variability of materials properties is now available in new materials as a result of research in the areas of synthesis and processing and of properties, offering designers almost unlimited variability in design choices. Often, use of such new materials is limited by the lack of evaluated design data on their properties.

- The performance of materials in actual systems depends on their response to the combined stimuli of stress in various forms, including mechanical and electromagnetic stress. The integration of understanding about materials performance and the design of devices and systems for full life cycles is an emerging phenomenon in industry that needs to be reflected in programs at universities and federal laboratories—where only limited activities can be identified today.

- Instrumentation for characterization and processing of materials is an increasingly important issue in materials science and engineering. Replacement of aging equipment, acquisition of new equipment, particularly for research on process technology, and funding for research on and development of new equipment are areas that deserve increased attention by funding agencies.

- Analysis and modeling find application in each of the four elements of materials science and engineering. High-speed computation has ushered in a new era in the use of these techniques, from the design of new materials to their ultimate use in products. Significant improvements in quality, reliability, and economy are the promise offered by increased application of analysis and modeling to all phases of materials science and engineering.

5

Manpower and Education in
Materials Science and Engineering

The interdisciplinary aspect of materials science and engineering is a critical, indispensable component of the field. Interdisciplinary interactions are so important that many industrial materials research laboratories and academic departments now comprise individuals from different disciplines who may work as members of these groups without ever fully losing their separate disciplinary ties.

Some of the great materials discoveries of history have been made by scientists, by engineers, or by craftsmen; many have been made by teams of all three types of individuals. The focus on materials has united the efforts of all three groups as they have sought to use materials practically and economically. Within universities, much emphasis is now being placed on improving interactions between scientists and engineers and on forging ties with industry to facilitate interactions with practicing engineers and craftsmen.

This chapter considers the educational problems and challenges posed by materials science and engineering. A brief description of the number and types of personnel employed in materials science and engineering is followed by an assessment of and recommendations regarding undergraduate, graduate, and continuing education. Also briefly discussed are issues related to precollege education and to the role of professional societies in promoting the development of materials science and engineering.

The field ranks high on the list of top careers for tomorrow's scientists and engineers. In a recent opinion survey, materials development was ranked as the most promising career path for young engineers (*Graduating Engineer*, March 1988). Experts questioned included the deans of engineering schools, school placement directors for engineering, and heads of firms that emphasize advanced technology and employ significant numbers of engineers. Every expert interviewed put materials research and development at or near the top of the list. Burgeoning needs were found in areas ranging from high-performance, specialty applications to high-volume mass production.

A related development involves the current efforts of physics and chemistry departments at many major research universities to expand their faculties in materials-related areas in response to the general perception of expanding opportunities in this field.

PERSONNEL IN MATERIALS SCIENCE AND ENGINEERING

The rich diversity of materials science and engineering is reflected in the wide variety of educational backgrounds represented by materials science and engineering practitioners (Table 5.1). Professionals who work as materials scientists or engineers include not only individuals with degrees from materials-designated departments (including materials science and engineering, metallurgy, ceramics, and polymer departments), but also individuals with degrees in chemistry, physics, engineering, and a wide range of other disciplines. The pluralism of its talented constituencies is a major contributor to the strength and promise of materials science and engineering.

The multidisciplinary nature of materials science and engineering complicates any assessment of personnel levels in the field, but rough estimates are possible (Figure 5.1). According to a report issued by the National Science Foundation (NSF), *U.S. Scientists and Engineers: 1986* (Surveys of Science Resources Series, NSF 87-322, NSF, Washington, D.C., 1987), there were 53,100 individuals employed in the United States in 1986 who identified themselves as materials scientists and engineers and had backgrounds in metallurgy, materials, or ceramics. There were also 72,600 physicists and astronomers and 184,700 chemists employed in the United States in 1986 (Table 5.2). By analyzing the subdisciplines of physics and chemistry, the

TABLE 5.1 Educational Backgrounds of Employed Scientists and Engineers in the United States

Profession	Doctorate	Masters	Bachelors	Others
Physicists and astronomers	56	21	23	—
Chemists	35	18	46	1
Materials engineers	13	27	51	9
All engineers	4	23	62	11

SOURCE: *U.S. Scientists and Engineers: 1986*, Surveys of Science Resources Series, NSF 87-322, National Science Foundation, Washington, D.C., 1987.

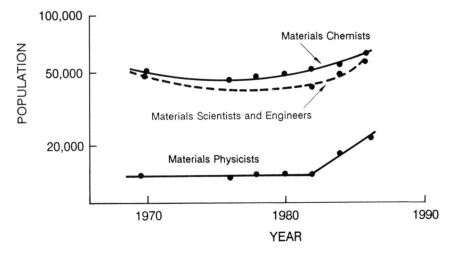

FIGURE 5.1   Estimated numbers of physicists, engineers, and chemists working in the field of materials science and engineering, 1970 to 1986.

TABLE 5.2   Scientists and Engineers Employed in the United States in 1986

Profession	No. Employed in U.S.	No. with Materials Focus
Physicists and astronomers	72,600	21,800
Chemists	184,700	61,000
Materials engineers	53,100	53,000
Nonmaterials engineers	2,390,000	?
Life scientists	412,000	?
Total		135,900

SOURCE: *U.S. Scientists and Engineers: 1986*, Surveys of Science Resources Series, NSF 87-322, National Science Foundation, Washington, D.C., 1987.

committee has estimated that 30 percent of the former group and 33 percent of the latter group have specialized in materials science and engineering. Therefore, for the purposes of this analysis, 21,800 materials physicists and 61,000 materials chemists can be considered to have been working in the field of materials science and engineering in 1986. By combining these estimates, the committee concluded that a core population of approximately 136,000 individuals was involved in materials science and engineering work in 1986.

TABLE 5.3 Primary Work Activity of Employed Scientists and Engineers in the United States (in percent)

Profession	R&D	Management and Administration	Teaching	Production and Inspection	Other
Physicists and astronomers	43	28	21	2	6
Chemists	40	25	14	15	6
Materials engineers	39	29	3	20	9
All engineers	33	30	2	17	18

SOURCE: *U.S. Scientists and Engineers: 1986*, Surveys of Science Resources Series, NSF 87-322, National Science Foundation, Washington, D.C., 1987.

Many engineers other than materials engineers regularly use materials or encounter materials-related problems. For instance, engineers who design electronic devices or are involved in aspects of their assembly regularly confront complex materials fabrication problems. However, a relatively small proportion of such groups is involved daily; most of the individuals are using the fruits of, rather than contributing to, materials science and engineering. Within the large and expanding population of life scientists, some individuals specialize in biomaterials. There is no way at present to estimate the contributions to the materials science and engineering community from these disciplines, so contributions from these groups also were not included in this analysis.

An important feature of the personnel statistics arises in their portrayal of work activities. Among materials engineers, nearly twice as many individuals work in R&D as in production or inspection (Table 5.3), and the relative proportions are much greater for materials chemists and physicists. This is another indication of the relative lack of emphasis accorded synthesis and processing within materials science and engineering. Changes in educational programs and industrial management will be necessary to involve more materials scientists and engineers in production-related activities. These changes must involve raising the perception of the intellectual level and value of these areas.

DEGREE PRODUCTION IN MATERIALS-RELATED DISCIPLINES

An important gauge of the interest and growth in materials science and engineering is the number of degrees granted each year in materials-related disciplines. As shown in Figure 5.2, the annual production of B.S. degrees from materials-designated departments was about 1000 per year in 1970; that number declined in the middle to late 1970s but now again stands at about 1000 per year. Figure 5.2 also shows the annual B.S. production rates in

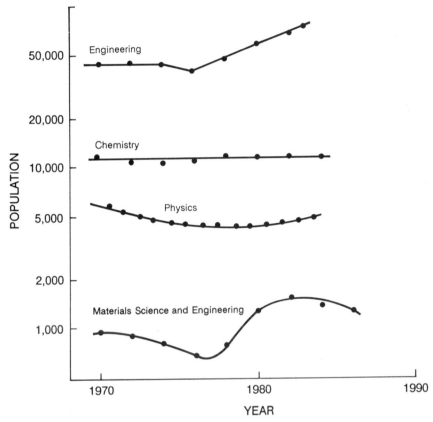

FIGURE 5.2 Estimated number of B.S. degrees earned annually from materials-related departments in U.S. universities, 1970 to 1986.

physics, chemistry, and engineering; the rate for engineering is about 50 times the production rate of bachelor's degrees from materials-designated departments. It is clear that the materials-designated engineering departments have not shared in the recent large increase in undergraduate engineering enrollment. The committee has concluded that this lack of growth is due to a troubling and continued lack of awareness of the field in high schools rather than to a lack of career opportunities (Table 5.4). A specific recommendation is to encourage changes in high school chemistry and physics curricula in order to introduce students to solid-state behavior and phenomena. Visible recent advances in superconductivity and magnetism should be exploited to achieve this goal. The flat enrollment picture may, in addition, mean that some materials-designated engineering departments are not capitalizing on the growth and expansion of the field.

TABLE 5.4   Degree Intentions of College-Bound High
School Seniors (percentage of total population)

Engineering Field	1981	1985
Chemical	0.7	0.6
Civil	0.6	0.5
Electrical	2.1	2.6
Mechanical	1.2	1.1
Metallurgy	0.04	0.02
Materials	0.009	0.01

NOTE: Based on a population of about 1 million seniors who reported
their academic plans.

SOURCE: The College Board, 45 Columbus Avenue, New York, N.Y.

The number of doctorates granted in materials-related specialties gives an indication of the number of new research-oriented practitioners entering the field. However, it is difficult to decide which of these specialties should be included within the field of materials science and engineering. Graduate work focused on polymers is conducted in chemistry and chemical engineering departments as well as in polymer and materials departments. Condensed-matter physics, a subfield of physics that now accounts for about 30 percent of all physics Ph.D. students, is a spawning ground for many who later work as materials scientists or engineers. Electronic materials are becoming an increasingly important concern of electrical engineering departments. An increasing number of civil, mechanical, nuclear, and aeronautical engineering departments have faculty and students whose primary focus is materials and materials issues.

For the purposes of this analysis, the committee chose to distinguish between subdisciplines within the core of materials science and engineering and those in which materials problems are an important subset of a broader field of interest. These core specialties consist of solid-state and polymer physics, polymer chemistry, and the engineering fields of materials science and engineering, metallurgy, ceramics, and polymers. Within these core specialties, virtually every doctoral candidate can be considered to be involved with materials science and engineering. A more expansive but less realistic view of materials science and engineering would include significant fractions of many engineering fields and physical, inorganic, organic, and analytical chemistry. Within these specialties, certain percentages of doctoral candidates (percentages that are not known accurately at present) can be considered to be working in materials science and engineering, but as a secondary interest.

In 1985, materials-designated engineering departments granted 343 Ph.D.'s, polymer chemistry departments granted 84 in polymer areas, and solid-state and polymer physics departments granted 259. When contrasted with the numbers from 1972, these figures reflect a growing number of Ph.D.'s being granted in polymer chemistry, polymer physics, and materials science and engineering, and a declining number in solid-state physics and metallurgy. The total of approximately 700 Ph.D. degrees awarded by these departments was little changed from that awarded 13 years earlier, in 1972. In 1985 the 343 doctorates granted in materials-designated departments equaled about one-half the number of physics doctorates and one-fifth the number of chemistry doctorates awarded in that year. The committee believes that students in the materials-related departments who focused on materials in earning their degrees will be a rich source of materials science and engineering professionals, given the growing importance of materials-related problems in many of these areas.

UNDERGRADUATE EDUCATION IN MATERIALS SCIENCE AND ENGINEERING

The committee believes that bachelor's programs leading to careers in materials science and engineering should serve two purposes: (1) they should provide sufficient grounding for a graduate to perform effectively over time in industry and (2) they should provide the fundamental underpinnings that will permit students who so desire to pursue graduate work in the field. As noted above, the majority of materials scientists and engineers do not earn advanced degrees, giving undergraduate education a special importance.

Whether or not a bachelor's student goes on to graduate school, designing course content on the basis of the present pressures of the marketplace is seldom wise, nor is it advisable that much of the content be devoted to immediate applications. Such expertise is generally short lived and creates quick obsolescence, particularly in a field developing as rapidly as materials science and engineering.

Undergraduate education can be divided logically into degree-granting and degree-supporting efforts. The former fit naturally into materials-designated departments. The latter can take a variety of forms, ranging from service courses in materials science and engineering to materials-based options in chemistry, physics, or engineering, to joint programs between two or more departments.

The committee recommends that, regardless of their institutional location and organization, undergraduate courses and programs in materials science and engineering be centered on the four basic elements of materials science and engineering—synthesis and processing, structure, properties, and performance—and on the relationships among them. The emphasis, where possible, should be on learning and applying fundamentals that cut across all classes of materials to ensure that students and researchers appreciate and understand why different materials compete with one another. New subjects should be developed that deal in a fundamental way with the control of structure through synthesis and processing to achieve desired structures and properties in materials at acceptable economic and social costs. A particularly compelling need is to provide a grounding in processing science and in the engineering of materials, and in how these relate to manufacturing processes.

The generic approach to materials science and engineering requires integrating disciplines and drawing contributions from physics, chemistry, and various engineering specialties. Obviously, universities should not simply teach separate courses in physical metallurgy, ceramics, and polymers and expect students to develop an integrated understanding of materials science and engineering. A sequence of generic courses on the synthesis and processing, structure, properties, and performance of all materials should be a central part of the undergraduate academic program of materials science and engineering departments. The most effective materials scientists and engineers will be well versed in fundamental principles and yet conversant with the general language of the multiple disciplines subsumed by the field.

Although the idea of integration has been repeated often in this report, its execution may be especially difficult at the undergraduate level, given the wide range of student interests and the constraints imposed by the goals of a liberal arts education. For this reason, the relative contributions of each field to the mastery of principles of materials science and engineering must be clarified and then integrated into the undergraduate curriculum. Some suggestions and models for accomplishing this were presented in a Materials Research Society Symposium on Materials Education (*MRS Bulletin*, May/ June 15, 1987).

Materials-Designated Departments

There are slightly more than 100 academic departments in the United States that offer materials-designated degrees at the bachelor's level. Increasing recognition of materials science and engineering as a broadly based field rather than a loose collection of disciplines is reflected in the growth of the number of those departments that identify themselves with the term *materials science and engineering*. In 1970 less than 50 percent of these departments had the term *materials* in their name, whereas today more than 80 percent do. Typically, the departments include courses taught by a faculty with expertise in metallurgy, ceramics, polymers, and perhaps electronic materials.

A few materials-designated departments have more focused interests, concentrating almost entirely on a single class of materials such as metals or polymers. In certain cases there are local or regional reasons why some materials departments should remain focused primarily on a single materials class. Particularly for schools with limited resources, the value of materials-specific programs should be recognized. But for most materials departments, a broader emphasis on materials is to be preferred.

The historical roots of materials departments are evident in the background and interests of the faculty now teaching in those departments. Of the estimated 1000 faculty members in materials departments in the United States, about 70 percent have metallurgy as a primary research focus. This group includes scientists and engineers specializing in extractive, processing, mechanical, and physical metallurgy, as well as those who have traditionally worked with metallic systems to understand general phenomena that can apply to many types of materials. Of the remaining 30 percent of the faculty members in materials departments, about 12 percent specialize in ceramics, 9 percent in polymers, and 9 percent in semiconducting, magnetic, or optical materials.

This profile of faculty in materials departments is hardly indicative of the tremendous diversity in the discipline. Apparently, the large and growing interest in materials other than metals has not yet manifested itself in a significant redistribution of faculty interests and expertise. Many of the faculty in materials departments who have specialized in metals are now nearing retirement age, so that these departments now have the opportunity to achieve a more balanced representation of materials classes.

The growing importance of materials classes other than metals could undoubtedly lead to a different distribution of research interest among materials department faculty in the future. Already, many former metallurgists and chemists are broadening their research and teaching and are pioneering in the development of the new field of materials science and engineering.

In the future, the hiring policies of materials-designated departments, particularly materials science and engineering departments, should reflect the goal of a comprehensive, generic educational program. This can best be accomplished by achieving a faculty balanced in background and research interests and of sufficient size to adequately address the burgeoning materials area, and by seeking faculty members whose interests extend to more than one materials class. Of necessity, smaller programs must create effective links with other departments to ensure adequate coverage.

Materials departments have made great progress toward the goal of broad-based programs in the years since the study by the NRC's Committee on Science and Materials Technology (COSMAT; *Materials and Man's Needs*, National Academy of Sciences, Washington, D.C., 1975) was issued. The substantial increase over the last decade in the number of departments describing their scope as materials science and engineering is one gauge of this progress. Another is the ongoing development of solid-state chemistry and physics courses for freshmen students in a number of materials departments, a development endorsed by the present committee. But old roots are strong roots, and most academic departments still have much to do to broaden and strengthen their undergraduate programs in materials science and engineering. Close links with other departments (e.g., polymers, chemistry, chemical engineering, and electrical engineering departments) will aid development of modern generic materials science and engineering subjects.

Role of Other Departments in Materials Science and Engineering

At most institutions, materials-designated departments do not account for the majority of materials-related activities, particularly in graduate education and research. According to an informal survey conducted by this committee, each faculty member in a materials department typically has at least two counterparts working on materials in other parts of the institution, such as in a chemistry, physics, or engineering department. Furthermore, most colleges and universities do not have materials departments, but many of these institutions have research and educational programs that can be classified as materials science and engineering programs. As was also concluded in the COSMAT study, accurate data describing current activities of faculty members outside materials departments are nearly impossible to obtain, but it seems reasonable to assume that the interests of this broad group are in areas encompassing the use of metals, ceramics, and polymers.

For example, the curricula of most university engineering departments include a course in materials, usually tailored to the specific interests of the specialty. In civil and mechanical engineering departments, students study a selection of materials on the basis of required design performance and life cycle costs. Similar subject matter is taught in aeronautical and nuclear engineering departments, with an emphasis on aspects of materials important to those fields. In electrical engineering, undergraduates are typically exposed to topics such as the electrical, optical, and magnetic properties of materials and the fundamentals of thin-film processing. Such courses are sometimes taught by faculty within the individual departments, sometimes by faculty from materials departments, and sometimes jointly by faculty from both departments.

These courses are a valuable part of engineering education, but a danger is that their focus may be narrow or outdated by the rapid development in the field. Faculty members who are educated in the fundamentals of materials science and engineering and are working at the forefront of the field have the advantage that they can adjust course contents according to advancements in the science and technology of materials.

One approach to the education of nonmaterials science and engineering students in materials is the joint teaching of subjects by faculty from materials and nonmaterials science and engineering departments. A second approach is to teach such students a sequence of two or more courses, the first given by materials science and engineering faculty and the second by nonmaterials science and engineering faculty. The first course should attend to the underlying principles of materials science and engineering. This course must lay the foundation for the second course, which should focus on the materials problems and needs specific to a given engineering field. For instance, mechanical engineers might have a second course that emphasizes the elastic-plastic properties of materials; chemical engineers might study the degradation, corrosion, and processing of materials; and civil engineers might study construction materials, composites, and so on.

Some materials science and engineering departments grew out of physics and chemistry. Now, many physics and chemistry departments are including components of materials science and engineering in their undergraduate and graduate curricula. This active development of discipline-specific courses focused on materials is a welcome and needed action. It appears likely, for example, that some chemistry departments will adopt curricular changes so that they can offer courses in synthesis and chemical preparation of materials, in the chemistry of materials processes, and ideally, in how synthesis and processing can affect properties in the design and manufacture of structures and devices. If successful, such redirection can serve as a model for other fields and disciplines that may wish to expand their efforts to emphasize materials.

Diversity and Integration

The preceding description of the dispersed institutional composition of materials science and engineering draws attention to the need for integration in the field. The committee believes that undergraduate materials engineering education should be centered in materials departments offering accredited degree programs with specialties in metallurgy, ceramics, or polymers, or broader programs that cover materials more generally. These departments should develop strong interactions in research and teaching with other engineering departments and with some science departments. Examples of such interactions include joint development and teaching of subjects, coordinated sequential teaching, joint faculty appointments, joint research activities, and interdepartmental degree-granting programs.

The specific outlines of such interactions will vary according to the institution, but the important feature is the construction of an undergraduate materials program around existing or developed materials departments or programs. The success of such ventures and the ability to clarify and exploit the relationships among materials, physics, chemistry, and engineering departments will in large measure determine whether the materials science and engineering field will be identified as having a major impact on the future health of R&D and of manufacturing in the United States. Such issues are currently being addressed, examined, and debated by such groups as the University Materials Council, representing heads of accredited materials science and engineering departments and of other materials-based programs, and key government and industrial leaders in the field.

Textbooks and Computers

An impediment to the development of broadly based undergraduate curricula has been the lack of good textbooks, particularly for intermediate-level courses. Although a number of excellent introductory texts are available, the paucity of more advanced textbooks and teaching materials is especially acute. Of the few intermediate-level texts available, hardly any attempt to integrate principles across materials categories, and none deals broadly with materials synthesis and processing. The lack of textbooks is exacerbated in part by the vast research opportunities for most materials science and engineering faculty. It is clear that materials science and engineering faculty now spend a much higher fraction of their time on research and are writing fewer textbooks today than in the past two decades. This shortcoming needs to be explicitly addressed and corrected through the direct intervention and, if needed, the financial support of government agencies and professional societies. The materials community should identify incentives for writing textbooks and other teaching materials that address materials as a whole and that focus on the interrelationships among synthesis and processing, structure, properties, and performance. These textbooks should also explicitly address the complementary approaches of physics, chemistry, and engineering.

A related deficiency—one shared with other areas of undergraduate education—is the underutilization of computers as teaching tools. Personal computers are now quite common on college campuses, presenting a tremendous opportunity for materials education. The materials community should develop programs that demonstrate and integrate the field's underlying principles. Here, too, incentives are needed to accomplish this worthwhile task. As has been shown in other fields, agencies such as NSF are ideally positioned to provide such opportunities through programmatic identification and support.

The Laboratory

An academic program that does not go beyond generic fundamentals would be a stale program indeed. An essential component of a student's education is applying these principles to real materials in the laboratory. Through experimentation, students can evaluate the validity of theories, gain familiarity with the tools of their future professions, and explore the interrelationships among the synthesis and processing, structure, properties, and performance of materials. These experiences are of vital importance to the modern materials curriculum.

At a minimum, a student's first laboratory course should deal with materials as a whole rather than with a single class of materials. Whenever possible, laboratory courses should use examples from several materials classes. One such course could focus on how the structure of materials determines their properties. The principles learned in these exercises could be applied in a subsequent laboratory course that emphasizes the synthesis and processing of materials.

The teaching value of these exercises depends greatly on the quality of the instrumentation in the laboratory. As discussed in Chapter 4, the tools of materials science and engineering are becoming increasingly expensive, making it difficult for institutions to equip their laboratories with modern instruments. An x-ray analyzer or scanning electron microscope far less sophisticated than the ones required for advanced research can cost more than $100,000. Equally expensive are the tools of modern materials processing and those for measuring the properties and performance of materials. Added to this is the great cost of computers and the interfaces and software needed to link instruments and computers.

Undergraduate laboratories at most universities are woefully inadequate, a consequence of the high cost of equipment and its maintenance. Ways must be found to upgrade undergraduate laboratories in materials science and engineering. The minimum cost for a basic materials laboratory is on the order of several million dollars. Keeping the laboratory up to date by replacing outmoded equipment requires annual funding equivalent to 10 to 15 percent of the initial investment. Solutions to this problem may have to include combining graduate and undergraduate laboratories and facilities, particularly expensive and often specialized instruments for structural characterization. The success of such an approach would depend on explicit ongoing support for technicians, maintenance, and upgrading.

Cooperative Programs with Industry

Besides working in the laboratory, students should be exposed to real materials and materials-related problems in other ways. Many universities now offer undergraduate students the opportunity to work with graduate students and faculty members on research problems. Cooperative programs with industry, such as summer jobs and work-study programs, are also becoming more common, exposing students to equipment, materials, and ideas that may not be available at academic institutions. Some of these programs have been broadened to allow students to do research for their senior theses, or to design projects, in an industrial setting. These programs should be expanded. In addition, a senior thesis or design problem—a proven means of exposing students to real materials-related problems and processes—should be a requirement incorporated into all materials science and engineering curricula.

GRADUATE EDUCATION IN MATERIALS SCIENCE AND ENGINEERING

The committee believes that the best materials scientists and engineers are well grounded in a core, materials-related discipline but are knowledgeable and respectful of the goals, philosophy, and methodology of several related disciplines. Focused project work (as is required in most master's degree programs) and thesis research (requisite for the Ph.D.) are proven means of encouraging specialization. Both serve as an apprenticeship in how to do research and how to apply theoretical principles. In addition, research at the doctoral level ideally should result in an original contribution to the field.

If students are to develop an appreciation for the productive interplay among disciplines, their graduate school experience must truly embody the interdisciplinary nature of materials science and engineering. Their education should demonstrate the fertile overlap of the chemical synthesis point of view of chemistry, the desire for a rigorous theoretical understanding of phenomena characteristic of physics, and the practical focus of engineering on end use. The aim should be to integrate a student's area of specialization—whether it be ceramics, condensed-matter physics, or materials chemistry—into the whole of materials science and engineering. Successful integration will breed familiarity with other disciplines, equipping students with the knowledge and skills they will need to exploit advances outside their specialty. In addition, experience in team research is particularly desirable and often obligatory in the more applied areas, since that is the common mode of applied work in industry. Such experience is also desirable in basic research on complex, interdisciplinary materials problems, as is well illustrated in research on high-T_c oxide superconductivity.

At the graduate level, as at the undergraduate level, materials science and engineering is at a point in its development where a growing fraction of the subjects taught to graduate students can be generic, dealing with materials as a class and with examples chosen from all materials classes. Among the courses that should be treated in this way are the thermodynamics, kinetics, properties, and processing of materials.

Of the four basic elements of materials science and engineering, the element of synthesis and processing has suffered the most neglect, both at universities and in industry. Universities should play a role in correcting this national weakness. Students who acquire understanding of synthesis and processing and an appreciation of problems encountered at the design and engineering end of materials development will bolster the ranks of scientists and engineers working to develop low-cost, high-quality materials processing methods.

Mechanisms for Achieving Goals in Graduate Education

Goals can be accomplished only with a concrete plan of action, the specific details of which will depend on the institution. But this committee believes that materials are so central to the development of modern science and technology, and to problems of productivity and international competitiveness, that each university should have the strengthening and broadening of materials science and engineering as important institutional goals.

Among the well-established means for fostering a productive mix of interests and activities are shared central facilities, courses taken in common and perhaps jointly taught and based on examples from all materials classes, interdepartmental thesis committees, shared use of teaching assistants, and seminars taken in common. At a minimum, universities should identify the full complement of courses that pertain to materials science and engineering. Faculty members interested in materials science and engineering should bear this responsibility. In some cases, cross-listing of courses in course schedules may be sufficient. Faculty members, however, must also help students select courses that not only pertain to their interests but also broaden their view of materials science and engineering. Core departments can further this aim by requiring students to take courses in other materials-related departments.

An option for small universities, where all of the core subdisciplines may not be represented, is to develop a multidisciplinary materials science and engineering program. Although not based in a department, such a program uses existing courses as a nucleus. New courses also may be required to achieve the breadth desired in materials education at the graduate level. High-quality multidisciplinary programs are notoriously difficult to maintain over long periods. Yet this approach may offer the best solution for universities that do not have a materials-related department with sufficient staff and resources to address the needs of graduate education in materials science and engineering. Dedicated faculty members are the key ingredient in making nondepartmental programs successful. Establishing a complementary research program—perhaps one that focuses on an inherently interdisciplinary area, such as polymer-based composites or biomaterials—may help to sustain a high level of faculty commitment.

Dedicated faculty members are also the hallmark of successful interdisciplinary and multidisciplinary research projects. Projects may run the gamut from joint research projects between two faculty members to large efforts carried out within the various collaborative entities sponsored by, for example, NSF or the Department of Defense. Interdisciplinary research programs that involve students are the most common means of their receiving substantial exposure to other disciplines. These interdisciplinary programs also provide faculty with a broader view of the field and with the impetus needed for developing broader and more generic curricula for their students.

An expected but not yet fully realized benefit of sponsored centers such as engineering research centers and materials research laboratories and groups is that they encourage collaborations with industry. Such collaborations allow graduate students working at these centers to experience how industry addresses materials problems, which usually require a multidisciplinary approach. Participating university faculty members should ensure, however, that this apprenticeship in how to do industry-oriented research does not slight the more academic educational aspects that should be incorporated into collaborative projects.

Examples of Institutional Arrangements

Implementing an interdisciplinary graduate education program requires unifying diverse materials communities and sometimes necessitates making formal arrangements that span long-standing divisions between departments, schools, and colleges. There are many alternative organizational models for accomplishing this. A large materials department can include faculty members who represent most or all of the core subdisciplines in the field, and its size can enhance its ability to develop ties with other departments.

In large or small departments, faculty members who hold appointments in two or more materials-related departments can serve as catalysts for interdisciplinary graduate education programs. Through their research, these faculty members are likely to expose students to the principles and detailed methodologies of several disciplines, as well as to the benefits of merging materials-related interests.

Some institutions do not have a department of sufficient size and with the necessary integration of disciplines to serve as a home for graduate programs in materials science and engineering. For these universities, one option is to create a school, institute, or other organizational division that performs this role. Regardless of its designation, the organization should be accorded equal status with other graduate divisions at the institution.

Whatever the specific approach to graduate education in materials science and engineering taken by a particular university, the implications of annual B.S. degree production in the United States (see Figure 5.2) must be recognized: there are simply not enough materials science and engineering graduates at the B.S. level to fill the current and projected needs for advanced degree holders in the field. Substantial numbers of students with undergraduate degrees in physics, chemistry, and other engineering fields must be encouraged to do graduate work in materials-related areas. New advanced-level introductory graduate courses must be developed to rapidly educate students who have not had at least a basic introduction to materials science and engineering at the undergraduate level. There is also a major need for texts that introduce principles of materials science and engineering at an advanced level to students who have been well grounded in scientific and engineering principles but not necessarily in materials science and engineering. Such texts would do for materials science and engineering graduate students what Kittel's text, *Introduction to Solid State Physics* (John Wiley & Sons, Inc., 1971), does for physicists who intend to explore condensed-matter physics.

CONTINUING EDUCATION IN MATERIALS SCIENCE AND ENGINEERING

The pace of technological change is now so rapid that the skills of scientists and engineers who do not stay abreast of new developments can become obsolete in only a few years. It is commonly held, for example, that engineers require some level of retraining at least every 5 years. In a dynamic, wide-ranging field like materials science and engineering, the risk of obsolescence and the consequent need for continuing education are especially great. Many companies, universities, and professional societies recognize the need for continuing education in materials science and engineering. Many options, varying in quality and effectiveness, are available.

Options for Providing Continuing Education

The traditional method of taking courses on campus works well for employees of companies located near a university. Most urban universities offer night classes, and many firms pay the tuition and fees of employees who take job-related courses. University-based short courses provide an intensity of focus and access to laboratory demonstrations and specialized on-site equipment.

Some large firms have integrated continuing education into their operations. Typically, they offer courses that are up to 15 weeks long, and subjects range from introductory physics to topics of technical interest to the company. In-house experts or university professors serve as instructors.

Less time-consuming than the preceding options, short courses and workshops that span one or several days are quite popular. National professional organizations, such as ASM International, the Materials Research Society, the American Physical Society, and the American Chemical Society, offer short courses before or after national conferences. In-house symposia and seminars are other means of keeping employees abreast of the latest technological developments.

The University of Minnesota, Stanford University, and the Illinois Institute of Technology are examples of academic institutions that provide live tele-casts of lecture courses to local firms. Telecommunications links allow stu-dents at remote sites to ask the instructor questions. Videotaped courses are an even more popular means of continuing education. The Association for Media-Based Continuing Education for Engineers (AMCEE), a consortium of 33 universities, offers a total of 518 courses in 16 disciplines, including materials science and engineering. This effort is supplemented by professional organizations that offer videotapes of short courses and workshops.

An outgrowth of AMCEE, the National Technological University (NTU) offers live and videotaped courses over a satellite network. NTU already offers a variety of courses in materials science and engineering, ranging from a short course on materials selection to live lectures from the University of Illinois course on composite materials for doctoral students, and it is con-sidering a master's degree program in the field.

Needs and Goals in Continuing Education

Industry and professional societies should take a more active role in en-suring that a multidisciplinary message is transmitted to materials scientists and engineers who need additional technical education or redirection and in strengthening educational efforts in materials synthesis and processing. More-over, it is recommended that the national associations track trends in materials science and engineering and identify developments that have the potential to increase the competitiveness of U.S. materials industries. This surveillance would help determine the content of continuing education programs.

Textbooks and other teaching aids are badly needed to help materials scientists and engineers grow with the discipline and to supplement video-taped courses and other instructional programs. One problem is that materials scientists and engineers constitute a small market. Textbook publishers thus have little incentive to address this need, and they require assistance from industrial and professional associations. The Educational Modules for Ma-terials Science and Engineering, a national body incorporated into the Ma-terials Education Council, has made a good beginning. It has prepared a sizable number of chapter-length teaching modules that have been published in the *Journal of Materials Education* but that unfortunately have not yet been widely adopted.

PRECOLLEGE EDUCATION

Although the committee has focused on undergraduate, graduate, and continuing education in materials science and engineering, it joins the many other bodies that have called for substantial improvement of science education in primary and secondary schools. Neglect of science at these levels imposes the burden of remedial education on university and college science and engineering departments. Perhaps more important, the nation's primary and secondary schools are failing to instill the awareness and curiosity that can inspire students to pursue a career in science or engineering.

For an area of great technological promise like materials science and engineering, this shortcoming can be especially severe. Current statistics on incoming engineering students suggest that only a few are interested in materials. This probably reflects the fact that the subject is usually ignored in high school chemistry and physics courses, which are the likely choices for an introduction to this fast-moving area of science and technology.

Materials scientists and engineers could well consider volunteering some time to their local school systems. Lectures on their careers, fields of science, or some aspect of materials science and engineering could provide vital stimulation, and such activities are greatly appreciated by school administrators and teachers. Industries, universities, government laboratories, and federal agencies could stimulate these activities by providing some sort of organization to help with contacts, provide guidance on quality, and advise on content.

Mentoring and tutoring in science-related topics are other activities that would be helpful. Summer employment and summer courses at universities also provide opportunities for precollege students to learn more about the nature of applied science and its role in the economy.

The professional societies, in particular, must seek ways to promote these activities. A greater variety and number of programs to help science teachers become more proficient in their fields and to more fully appreciate the role of science and technology in the economy could be organized. Programs such as summer jobs in industry and summer courses at universities need to be organized for teachers and should be sponsored by industry, universities, professional societies, and the states.

The professional societies, NSF, and other government agencies should make efforts to use television to work toward many of the goals described above. The Public Broadcasting Service, through series such as *Nova* and a wide variety of nature programs, appears to have had a significant impact on public attitudes toward basic science and environmental concerns. It seems that similar programs could be developed that would increase the prestige of manufacturing and production activities. It was not many years ago that American society glorified the activities of men like Edison and Ford. It is important to recapture some of those attitudes. The popularity of series like *Nova* and the number of magazines describing scientific advances for the general public show that the public is receptive.

ROLE OF PROFESSIONAL SOCIETIES

The many societies devoted to materials sciences and engineering, some of them recently established and others more than 100 years old, illustrate the field's richness and diversity. Some have as their focus specific materials such as metals, ceramics, plastics, and composites. Others began as divisions of already well-established societies in physics, chemistry, and engineering. Still others have their roots in materials research, engineering, and processing, and a fourth group sprang up around techniques and phenomena such as electron microscopy, crystal growth, corrosion, and testing. Almost all have broadened their initial foci as the materials science and engineering field has evolved. Most perform some or all of the functions described below.

Professional societies facilitate exchange of information at society-wide and regional meetings for presentation of experimental results, new theories, and review papers. The scope of such meetings ranges from brief research-in-progress presentations to in-depth symposia on specific technical topics. Dissemination is enhanced by the publication of journals, transactions, or magazines as well as conference proceedings. The societies also provide education on many different levels, including in-depth tutorials in emerging areas of the field; home study, training, and video courses; and special seminars. Some societies provide scholarships for college freshmen or upperclassmen to encourage study in the field.

The societies promote information gathering and dissemination, including collection and publication of data bases, bibliographies, abstracts, translations, handbooks, and other information services. Many of these are available as software programs for personal computers, and eventually they will be available as on-line services to subscribers. Other information may deal with government policy issues such as tax incentives, investment credits for R&D, patent policy, environmental policy, and international competitiveness.

Similar functions are performed by societies in other technical fields, but the professional societies in materials science and engineering seem particularly important because the field is still young, relies a great deal on dissemination of experimental research, and encompasses a very broad range of interests. Society activities also provide a common bond and opportunities for professional interactions among members and help to develop the interdisciplinarity of the field. Greater cooperation among the societies could be advantageous for advancing the field as a whole.

FINDINGS

- The field of materials science and engineering is pluralistic, drawing significant numbers of its practitioners, particularly at the Ph.D. level, from physics, chemistry, and allied engineering fields, as well as from materials-designated departments.
- Undergraduate courses and programs, regardless of departmental location, should emphasize the four basic elements of the field and their in-

terrelationships: synthesis and processing, structure, properties, and performance of all classes of materials. The area of synthesis and processing has suffered neglect in our universities and in industry. A particularly compelling need is to provide undergraduates with a thorough grounding in the science of the engineering of processing and its relation to manufacturing. At the graduate level, students exposed to synthesis and processing research activities will be better equipped to contribute (and more interested in contributing) to this area of industrial need.

• New courses and new textbooks, dealing generically with all materials, are needed at both the undergraduate and graduate levels. A special need is evident in the area of synthesis and processing, covering the spectrum from processing science to manufacturing.

• Undergraduate materials engineering education should be centered in materials departments. Such departments should interact strongly with other science and engineering departments to develop interdisciplinary materials-related educational programs.

• Graduate education in materials science and engineering should emphasize a sound education in a specific discipline, while providing understanding of, involvement with, and respect for the goals, philosophies, methodologies, and tools of complementary disciplines.

• Existing institutions for continuing education should be expanded and should be more widely publicized, and badly needed appropriate textbooks should be produced. Assistance and encouragement from industrial and professional associations are critical to both endeavors.

• The many professional societies associated with materials sciences and engineering should establish mechanisms for cooperative action to advance the field as a whole.

• Perhaps most importantly, the annual production of bachelor's degrees for materials-related departments is currently about 1000 per year, a figure that has changed little from the early 1970s. The number of doctorates is just under 700, again a figure little changed from the early 1970s. Thus the production of educated professionals has remained essentially constant in the face of greatly increased needs and opportunities in the field and expansion of the field to include new materials such as electrooptical materials, advanced composites, and high-temperature superconductors. Additional educated personnel are required to meet these needs and opportunities. National strengthening of synthesis and processing and of performance will require additional manpower.

6

Resources for Research in Materials Science and Engineering

The major resource needed for research in materials science and engineering (and indeed for any research program) is trained scientific and engineering manpower. The education of materials science and engineering researchers has been considered in Chapter 5. This chapter describes the institutions and funding that support the acquisition and operation of the equipment needed for research in materials science and engineering. This chapter also describes the research environment for the scientists and engineers who use this equipment.

Research in materials science and engineering differs from that in other fields because it spans the full spectrum from basic research to practical applications. There is strong industrial commitment to such research, especially applied research. In addition, government makes funds available to university and industry researchers, and it provides research support for materials science and engineering at its own laboratories, where it supports basic research and research for technologies (e.g., weapons development or nuclear energy, areas in which, by policy decision, the government plays a primary role). It also makes large, unique facilities available to the industrial and university research communities.

The scale of the organizational structure for materials science and engineering research ranges from individual researchers through small organized groups that carry out basic research programs to large compartmented research efforts in development and manufacturing. The equipment used—including lasers, electron microscopes, molecular beam epitaxy systems, high-field magnets, neutron beam reactors, and synchrotron light sources—varies greatly in size, cost, and complexity of operation, but the fundamental nature of the research organization does not change profoundly with equipment requirements.

Research in materials science and engineering is carried out predominantly in the "small science" mode, in which groups of a few researchers attack several aspects of a problem. This is true of all materials science and engineering efforts in synthesis and processing, characterization, and the study of materials properties and performance. Even when the largest facilities, such as reactors (neutron sources) and synchrotron light sources, are used, as in structural characterization studies, the experiments are small-scale, and the individual experiments usually involve only one or two researchers. Research efforts at these facilities are large only in the sense that many experiments can be carried out at one time.

It is therefore not particularly illuminating to make distinctions in types of materials science and engineering research based on the size of the equipment used. Decisions on allocation of resources should be based on resource requirements for attacking problems in the four basic elements of materials science and engineering. That materials science and engineering is a collection of disciplines (e.g., physics, chemistry, and metallurgy), and not a single discipline, makes it necessary to examine resources in a crosscutting way. It makes collecting data on resources more complex, as can be seen in the attempts even to estimate the amount of money being spent on materials science and engineering. In looking at large pieces of equipment (e.g., light sources), account must be taken of the fact that much of their use—for medical research, for example—is not related to materials science and engineering and that much of the work in materials science and engineering disciplines such as physics is not directly related to materials science and engineering. These facts argue strongly for allocating resources according to the requirements of problems rather than on the basis of the subdisciplines or the size of the equipment involved.

FEDERAL FUNDING FOR RESEARCH IN MATERIALS SCIENCE AND ENGINEERING

At least 15 federal agencies, each with different characteristics and missions, sponsor R&D in materials science and engineering. Most of these agencies do not break down their R&D activities into categories identifiable specifically as materials research or development. Furthermore, because no single definition for materials science and engineering is commonly accepted by these agencies, reports without analysis may be misleading. Thus it is very difficult to obtain an accurate estimate of the total amounts, much less a breakdown by agency or subfield, of funding provided by the federal government for materials science and engineering. Materials-related work can occur as part of systems development, as part of an exploratory program, or as basic research done for its own sake. Since materials lie at the heart of many technological systems, it is not easy to separate out the materials science and engineering part of a technical program. For instance, within a program to develop solar-electric technologies, materials research may be required to understand and improve photovoltaic materials for converting photons to electrons. However, such research is done as part of a system, so the fraction of the total project dealing with materials science and engineering has to be estimated. This situation is exacerbated by security issues when Department of Defense (DOD) R&D is probed, and figures reported for R&D related to materials for that agency are always presumed to be lower than actual expenditures.

In view of these difficulties, special efforts are necessary to obtain an overview of the level of federal support for materials science and engineering. The history of these survey efforts reveals a lack of long-term continuity and a lack of consistency in criteria for inclusion of, for instance, the various expenditures associated with mineral location, refinement, and beneficiation. The efforts made during this study to extract some consistent data from these various surveys are summarized in Table 6.1 and are discussed below.

In 1977 the Committee on Materials (COMAT) of the Federal Coordinating Council for Science, Engineering and Technology, which is convened by the Office of Science and Technology Policy, released its *Inventory and Analysis of Materials Life Cycle Research and Development in the Federal Government*. The committee estimated that in FY1976, at least 15 agencies of the federal government spent a combined total of approximately $1 billion on materials research and technology (which it defined somewhat more broadly than R&D). Five agencies—the Department of Energy (DOE), the Department of the Interior (DOI), DOD, the National Science Foundation (NSF), and the National Aeronautics and Space Administration (NASA)—spent more than $100 million each. This survey and a follow-up study for FY1982 were carried out for COMAT by DOI and included all activities that could be broadly defined as related to the entire materials cycle, including minerals location, refining, and beneficiation, and all forms of waste disposal (including, for example, the extensive programs in the Environmental Protection Agency, which involve primarily analytical chemistry). This inclusion alone added $99 million to the materials R&D inventory. A similar study of FY1980 activity was developed for COMAT but was carried out by the Department of Commerce (DOC) using a narrower definition of activities closer to that emphasized in this current materials science and engineering study. Although differing somewhat in scope, these three surveys did share the methodology of defining materials science and engineering R&D activities with some precision and obtaining detailed responses from the departments and agencies queried. Total materials funding was reported, and it was also broken down into funding for basic, applied, and developmental work; by research end

TABLE 6.1   Materials R&D by Federal Agency (millions of dollars per fiscal year)

Agency	1976[a]	1980[b]	1982[c]	1986[d]	1987[d]
DOE	333	514	286	440[a]	440[e] (est.)
DOD	132	160	147	374	338
Bureau of Mines	24	23	30	27[f]	27[f]
NSF	69	89	100	133	138
NASA	52	79	101	84	128
NIST	21	36	14	23	26
Total	631	901	678	1081	1097
DOD	132	160	147	374	338
DOE (defense)	59	110	64	104[e]	104[e] (est.)
Total, reported defense	191	270	211	478	442
Total, nondefense (current dollars)	440	631	467	603	655
Federal science deflator[g]	0.604	0.834	1.000	1.145	1.166
Total, nondefense (constant 1982 dollars)	728	757	467	526	562

[a]Committee on Materials (COMAT) report, 1976.

[b]COMAT report, 1980.

[c]COMAT report, 1982.

[d]Federation of Materials Societies report, 1988.

[e]Private communication, L. Ianiello, DOE.

[f]Private communication, M. Schwartz, Bureau of Mines.

[g]"Implicit Price Deflator for Total Government Non-defense R&D," Bureau of Economic Analysis, Department of Commerce.

use (relating to defined national goals); by stage of materials cycle (from exploration of resources through extraction, processing, manufacturing, and fabrication and on to application and evaluation); by performer (university, industry, agency, other federal laboratories, and nonprofit laboratories); and by materials class.

It is interesting to note that in 1980, 90 percent of the total federal funding for materials R&D was provided by six agencies—DOE, DOD, DOI, NSF, NASA, and DOC—with very limited efforts by the other nine reporting sources. Funding reported for basic, applied, and developmental efforts totaled $326 million, $413 million, and $358 million, respectively. Of the FY1980 federal materials R&D budget, 32 percent was spent to improve energy supplies, 23 percent to improve national security, 16 percent to improve the science and technology base, and 5 percent to improve industrial productivity; 23 percent of the work was performed in-house, 30 percent in non-civil-service federal laboratories, and 42 percent by the private sector (23 percent by industry and 19 percent by universities).

It is regrettable that COMAT has not continued these surveys in succeeding years. Would it still have been true in 1988 that only 5 percent of materials R&D was aimed at improving industrial productivity—a question of considerable relevance to today's concerns about industrial competitiveness? Has the proportion of materials R&D associated with national security increased as markedly as total federal R&D? How has the total materials science and engineering effort fared in recent years? The questions are myriad, the data almost nonexistent, and the answers few.

Although no survey as detailed as that provided by COMAT is available, in recent years the Federation of Materials Societies (FMS) has assembled and published estimates of total federal R&D in materials science and engineering research for the agencies listed in Table 6.1. Their summaries for FY1986 and FY1987 are included in this table, with two modifications. The data presented for FY1986 and FY1987 reflect private communications with DOE and the Bureau of Mines to limit reporting to the same categories included in the COMAT studies. Data from the FMS suggest that FY1988 and proposed FY1989 budgets show only minor agency-by-agency adjustment in these years of budget stringency.

The results summarized in Table 6.1 reveal an increase in total federal funding for materials science and engineering by the six indicated departments and agencies from $631 million in FY1976 to $1097 million in FY1987. These numbers probably underestimate the total federal effort by about 10 percent, which is the percentage associated with work by other unlisted agencies, and by a large but unknown amount for unreported defense activities.

It is instructive to separate the activities associated with national security from the remainder of the effort. This has been done in Table 6.1, in which DOD and DOE (defense) expenditures for materials R&D are subtracted from the totals. The net nondefense expenditures were nearly constant over the 11-year period examined, with fluctuations primarily attributable to variations in DOE funding—which was up during the late 1970s to accommodate energy-related development programs, down in the early 1980s in response to the demise of most of those programs, and up once again in the middle to late 1980s to fund DOE waste disposal programs. Decreasing funding to the Bureau of Mines was roughly canceled by increasing funding to NSF.

Not unexpectedly, when the appropriate federal R&D deflator is applied to the data in Table 6.1, they show a 21 percent decrease in effective level of effort during the 11-year period surveyed. When the increased commitment by DOE and NSF to the support of major facilities for materials science and engineering research is incorporated into the analysis, it comes as no surprise that workers in this field have experienced a dramatic reduction in their combined level of effort. This reduction in level of effort coincides with a period of growing opportunity, as described throughout this report, and with a period of rapid advances in materials science and engineering by our economic competitors in the world market.

Three observations based on these data are particularly pertinent. First, the FY1988 federal budget includes approximately $100 million for research on high-temperature superconductivity, all but about $18 million reprogrammed from other efforts. The impact on other important areas of research must be substantial. Nearly 10 percent of the total federal effort in materials science and engineering is devoted to this one field. New, not reprogrammed, funding for high-temperature superconductivity is clearly called for.

Second, our international industrial competitors have all identified materials science and engineering, computer and information technology, and biotechnology as priority technologies and have initiated and funded major national R&D efforts to support the development of commercially successful enterprises in these areas. The data presented here reveal that, with the exception of the provision for national security, the United States has elected to decrease its level of effort in materials science and engineering.

Finally, the committee emphasizes once again the need for accurate data for carrying out rational analysis of this important field. The National Critical Materials Council, working with COMAT, is the logical leader to reinstitute a federal materials science and engineering inventory with a level of detail comparable to that of the 1976, 1980, and 1982 surveys.

Department of Energy

In FY1987 DOE reported spending almost $500 million for materials-related R&D. Funds for materials-related R&D associated with classified weapons programs were not included in the totals. The reported spending was distributed among six different activities: 30 percent in materials sciences, 30 percent in nuclear energy, 15 percent in magnetic fusion, 12 percent in conservation, 10 percent in solar energy, and 2 percent in fossil energy.

It is of interest to examine budgets within the Materials Sciences Division (MSD), because MSD funds most of the construction and operation of the large facilities for the national materials program. In FY1987 the MSD budget was $155 million in operating funds and $15.5 million in equipment funds. Breaking it down by recipient, 63 percent went to DOE laboratory programs, 35 percent went to university programs (including laboratories where graduate students are involved in research to a large extent, e.g., Lawrence Berkeley Laboratory and the Iowa State University Laboratory at Ames), and 2 percent went to industrial and other projects. Breaking it down another way, 22 percent went to facility operations, and 78 percent went to contract and grant research. Of the contract and grant funding, 64 percent was spent in metal-

TABLE 6.2   Number and Size of Grants from the Materials Sciences Division (FY1983 to FY1987)

Grants and Funding	1983	1984	1985	1986	1987
Number of grants	207	222	213	240	246
Grant funding ($ million)	22.1	26.2	27.7	28.6	33
Average grant size (real dollars)	107,000	118,000	130,000	119,000	134,000
Average grant size (1985 dollars)	121,000	124,000	130,000	113,000	119,000

SOURCE: Materials Sciences Division, Department of Energy.

lurgy, materials science, and ceramics; 27 percent went to solid-state science and solid-state physics; and 9 percent went to materials chemistry. The projected funding for FY1988 for MSD was $169 million. Since FY1984, funding both for universities and for major facilities has increased within the division, whereas support for research in the DOE laboratories has decreased.

Table 6.2 shows the number and size of the grants issued by MSD for FY1983 through FY1987. When these data are normalized to 1985 dollars, the corresponding average grant values become $121,000, $124,000, $130,000, $113,000, and $119,000, respectively. The decrease in grant size for FY1986 was due to the Gramm-Rudman-Hollings deficit reduction legislation.

During 1987 about $10 million was reprogrammed from ongoing MSD research programs into research on the new high-temperature superconductors. Throughout the federal government, reprogramming of funds for superconductivity research totaled approximately $50 million in FY1987. It is not clear what has been lost by the reprogramming; some funding undoubtedly came from low-temperature superconductivity programs, and some came from ceramics programs. It is estimated that about $100 million will have been spent on superconductivity in FY1988.

Congress has initiated the support of several new university campus buildings from within the MSD funding. This is very detrimental to ongoing programs, but the pressure for such initiatives is increasing.

Department of Defense

For FY1987 DOD reported a budget of $316 million for its Materials and Structures Science and Technology Program. Table 6.1 shows DOD as having spent $338 million in FY1987. The $100 million for Sematech was a separate DOD budget. Whatever the reported figures, it is understood that large amounts of funding for classified materials R&D go unreported.

The Materials and Structures Science and Technology Program is managed by the Air Force (50 percent), the Army (25 percent), the Navy (15 percent), and the Defense Advanced Research Projects Agency (10 percent). More than 80 percent of the work is conducted by industrial and university groups. Materials science and engineering work is also conducted in about 124 industrial laboratories under the independent research and development (IR&D) programs. These programs are not directly sponsored or required by a DOD contract or grant; the costs are reimbursed by DOD through overhead accounts. In total, DOD funding supports a significant portion of the nation's science and engineering base.

Programs funded by DOD are often directed toward specific applications. Table 6.3, from the Office of Research and Advanced Technology, lists some of the mission areas, technology needs, and thrusts of the DOD's Materials and Structures Science and Technology Program. In many cases, the mission areas are unique, the technological demands rigorous, and the costs irrelevant,

TABLE 6.3   DOD Materials and Structures Science and Technology Program

Mission Areas	Technology Needs[a]	Thrusts
Strategic offense Reentry vehicles, propulsion systems	All-weather capability, maneuvering capability, efficient rocket nozzles, lightweight upper stages	Carbon-carbon composites, metal matrix composites
Space Satellite structures, propulsion systems, mirrors and optical structures, antennas	Survivability, no outgassing, thermal/electrical conductivity, dimensional stability, high stiffness, damping	Metal matrix composites, ceramic matrix composites, carbon-carbon composites
Land warfare Tanks, vehicles, mobility	Improved armor, gun barrel erosion, ground vehicle survivability	Metals, ceramics, organics; metal matrix composites
Air warfare Aircraft, tactical missiles	Durability of composites, high-strength "forgiving" metals, long-life high-temperature gas turbine components, all-weather-capable seeker domes	Organic matrix composites, metal matrix composites, ceramic matrix composites
Naval warfare Mines and torpedoes, ship survivability, submarines	High-strength "forgiving" metals, composites, joining techniques	Metals, metal matrix composites, welding
Research	Understanding structural response, energy interactions, synthesis of new materials	Micromechanics and macromechanics, fracture mechanics

[a]Materials properties, loads and environments, characterization, nondestructive evaluation.

SOURCE: Office of Research and Advanced Technology.

so that program success produces a material of no use to the domestic market. A good example of such a "boutique" material is the recently announced tank armor made of steel-clad depleted uranium. Note that Table 6.3 does not indicate any technology needs in the areas of electronics or communications.

National Science Foundation

Like DOE, NSF prepares reports and summaries of its materials-related research. This research is conducted primarily through its Division of Materials Research and also through its Division of Chemical and Process Engineering and Division of Mechanical Engineering and Applied Mechanics.

The Division of Materials Research and the divisions of engineering within NSF seek to provide balanced support among three fields—materials engineering, materials chemistry, and condensed-matter physics. They also maintain balance among various funding modes—individual investigator projects, interdisciplinary groups of various sizes, centers for materials research, and national facilities. In addition, the divisions provide support for instrumentation. Funding for the divisions supporting materials work has dropped slightly in real terms over the past decade.

The impact of the NSF program is discussed later in this chapter, in the sections "Small Group Research" and "Collaborative Centers." What follows below are brief descriptions of NSF program funding for materials research laboratories, materials research groups, industry-university cooperative research centers, and engineering research centers.

The National Science Foundation supports nine materials research laboratories that had a combined budget of $26 million in FY1987. These collaborative centers in interdisciplinary materials science and engineering were established in 1960 by DOD through the Defense Advanced Research Projects Agency, and they were transferred to the NSF in 1972. By promoting coherent, interdisciplinary, multiinvestigator projects, these laboratories continue to make major contributions to the evolution of materials science and engineering.

Some 25 other universities have established materials research laboratories (under that or related names). They are supported by a variety of contracts and grants from federal agencies and industry. Their annual budgets range from $0.5 million to $8.0 million.

In 1985, NSF began a program to fund materials-related collaborations smaller than the materials research laboratories but larger than individual investigator projects. Known as materials research groups, these collaborations of 5 to 10 researchers are being supported with grants of around $0.5 million to $0.75 million per year. In the first five materials research groups formed, the topics under consideration included the stability of glass, the aging of polymer blends, new low-temperature ceramics, the fundamentals of photoelectrochemistry, and the interfaces between different solids and between solids and liquids.

Another important forum for collaborative research that is supported by NSF is the industry-university cooperative research center. Initiated in 1973, these centers bring together faculty and industrial researchers to work on problems of common interest. Funding comes both from NSF and from industry, and a center is considered a success when its industrial support covers both direct funding and equipment costs. Most centers level off at an annual level of about $500,000 to $1 million in nonfederal funding for each center. NSF support for these centers is phased out after 5 years. At present, there are more than 35 such centers devoted to materials science or related activities, with total funding of over $30 million.

Other recent examples of interdisciplinary centers supported by NSF are the engineering research centers that were initiated in 1985. By bringing into a single research structure individuals from a wide range of scientific and engineering fields, the engineering research centers promote the kind of interdisciplinary teamwork that emerging technological problems demand. There are currently 13 engineering research centers, 10 of which are directly related to materials or have potential materials applications. The existing centers are funded at an average level of $1 million to $3 million per center per year for an average of 5 years.

A final set of centers are those funded by the states. Some dozen of these centers, largely in the northeastern states, focus on materials and have budg ranging from $1 million to $4 million per year.

INDUSTRIAL FUNDING FOR MATERIALS
SCIENCE AND ENGINEERING

Overall in 1987, industry in the United States spent an estimated $60 billion on R&D—about the same amount as the total level of R&D funding by the federal government. This amount has nearly doubled in constant dollars over the past two decades. Industry also conducts about one-half of the R&D funded by the federal government. Industry therefore performs about three-quarters of the R&D done in the United States.

Many of the difficulties encountered in estimating federal support for materials science and engineering also arise in considering industrial funding of materials research and materials-related projects. For instance, industry reports on the total magnitude of its R&D expenditures through the 10-K forms submitted to the Securities and Exchange Commission, but rarely do companies break their total expenditures down into categories. Even if this information were available, it would be difficult to separate materials-related efforts from nonmaterials activities.

No recent attempt has been made to survey the level of materials science and engineering R&D in industry, and the committee has had to turn once again to the results of a COMAT survey carried out as Phase II of the *Inventory and Analysis of Materials Life Cycle Research and Development*. The document, published in 1979, is based on statistical data from U.S. industry for a 12-month period during 1977 and 1978 and should be roughly comparable with the Phase I study for government-funded R&D for FY1976. The methodology of the study was sophisticated: each industrial sector was sampled in a statistically significant survey that included detailed categorization of research by market classification, functional emphasis, national goals, stage in the materials life cycle, and materials classes. Survey results were extrapolated using industry-wide data for total R&D published annually in *Business Week* magazine; data were analyzed and compared on a category-by-category basis with federal expenditures for materials science and engineering R&D.

Not all industrial R&D was readily assignable to categories, and, consequently, totals differed significantly when summed by different categorizations; however, in general the COMAT survey revealed that the industrial expenditures for R&D were more than 4 times those of the federal government. For example, as shown in Table 6.4 for various stages in the materials life cycle, industrial expenditures of $2,950 million can be compared to the federal total expenditure of $653 million. These industrial data exclude R&D

TABLE 6.4   Industrial and Federal Support of Materials R&D for Various Stages in the Materials Life Cycle

Stage	Industrial Support		Federal Support		Total Support		Fraction Government
	$ Million	Percent	$ Million	Percent	$ Million	Percent	
Exploration and extraction	88	3	52	8	140	4	0.37
Production of finished material	326	11	26	4	352	10	0.07
Development of new materials	1402	48	327	50	1729	48	0.19
Development of new processes and applications	987	33	182	28	1169	32	0.15
Recovery and recycle	79	3	7	1	86	2	0.08
Waste management	68	2	59	9	127	4	0.46
Total	2950	100	653	100	3603	100	0.18

SOURCE: COMAT report, 1980.

related to fuels but are probably still rather liberal in defining materials science and engineering relative to the definitions of this current report.

While the COMAT survey is more than 11 years out of date, it is revealing in several ways. First and foremost is the fact that industrial research on materials science and engineering represents a much larger fraction of the total industrial R&D effort (19 percent of an estimated total of $19 billion) than is represented by federally funded research (5 percent of an estimated total of $21 billion). Coupled with the fact, revealed in Phase I of the COMAT study, that about 20 percent of federally funded R&D in materials science and engineering was performed by industry, is the fact that 83 percent of all materials science and engineering R&D was performed by industry in 1977. Extrapolation to today's budgets is fraught with danger, but it is notable that if a comparable 19 percent of industrial R&D in 1987 were devoted to materials science and engineering, it would represent a commitment in excess of $11 billion, an enormous sum of money in anyone's accounting scheme.

Another rough estimate of industrial funding for materials R&D can be derived from the personnel figures discussed in Chapter 5. According to NSF's *U.S. Scientists and Engineers: 1986*, 17,300 of the 45,200 materials engineers who worked in industry engaged in R&D as their primary work activity. Similarly, 10,800 of the nation's physicists and 54,900 of the nation's chemists worked predominantly on R&D within industry. By again using the very rough estimate that 30 percent of physicists and 33 percent of chemists are engaged in materials science and engineering, the committee estimates that 3,200 physicists and 18,100 chemists work on materials R&D within industry. The level of support for individual scientists and engineers working within industry varies considerably, but the committee estimates that $200,000 per person per year is a reasonable cost estimate to apply to the group as a whole. This estimate suggests that approximately $7.7 billion is currently spent on materials R&D within industry. However, not all of that amount is provided by industry, since some of industry's R&D on materials is supported through federal R&D contracts and grants.

It is certain that the figures reported for industry are much more heavily weighted to development (including product development) than are the federal expenditures, but no quantitative measure of that distinction was attempted by COMAT. It is interesting to note that the 1980 COMAT study estimated that basic research accounted for about one-third of federal expenditures for materials R&D. That would translate into about $200 million in FY1976. A comparable expenditure by industry in 1977 would have represented only 7 percent of the total R&D in materials science and engineering, not an unreasonable number if one acknowledges the huge efforts directed to basic research in materials science and engineering at the laboratories of such companies as Bell Telephone, IBM, General Electric, and Du Pont. It is probably quite imprecise to suggest, as many do, that most basic materials science and engineering research is done at universities and government laboratories.

Finally, in discussing trends in industrial R&D since the COMAT study of 1977, it is reasonable to conclude that industry has shifted increasingly toward applied and developmental research and away from basic research. The disappearance of many research laboratories in the metals industries is the most visible manifestation of this trend.

RESEARCH SETTINGS

The settings in which materials scientists and engineers work can be divided into two general categories according to their administrative structures and the amount of funds invested in equipment and instrumentation. The first category consists of small groups—for example, one or a few researchers, a small team, or a principal investigator and several postdoctoral, graduate, and undergraduate students—funded in the range of several hundred thousand dollars or more. The second consists of collaborative research centers funded at levels of a few million dollars or less, as exemplified by the materials research laboratories and engineering research centers supported by NSF and by state funding. An example of the kind of equipment that can be found at such centers is depicted in Figure 6.1, which shows a molecular beam epitaxy system used for preparation of mercury-cadmium-telluride artificially structured materials. In addition, there are major national facilities, such as synchrotron radiation and pulsed neutron sources, that are funded at levels of many millions of dollars—often over $100 million. The research at these facilities is carried out by small groups.

The lines of demarcation between these categories are far from clear. Individual investigators oversee small group research both in collaborative research centers and in major national facilities. There are also many smaller national facilities that serve user communities in the same way as the major facilities do. But these types of research environments can readily be distinguished, and each has capabilities and problems unique to that setting.

Small Group Research

Much of the progress in materials science and engineering has been inseparably associated with small groups of researchers headed by an individual investigator. To take one example, the materials-related work awarded the Nobel Prizes in physics from 1985 through 1987 was all done in small groups. Several of these prize winners have also made use of large facilities in their work. Such groups are common throughout the field—in industry, in private research institutions, in government laboratories, and in universities. Typi-

FIGURE 6.1 Molecular beam epitaxy system capable of growing thin films and multi-layer structures in ultrahigh vacuum using Knudsen sources. Modular construction provides substrate loading and preparation capabilities and in situ film analysis. (Courtesy Varian Corporation.)

cally, groups in different sectors have different goals: research in industry is more oriented toward products and services, in federal laboratories toward particular missions, and in universities toward fundamental understanding. But the modes of operation in each sector are similar.

Stiff competition for research support has forced investigators to focus on short-term projects rather than risk having too little progress to report in renewal proposals. High-risk speculative research usually correlates with major advances in a field. However, results from such research, if they occur at all, often require longer to achieve than the usual 1- to 3-year funding period. If there is a long period of support without results, continuation proposals refereed by peer groups are likely to receive low marks, leading to reduced levels or loss of support. Difficult and unconventional interdisciplinary research, which is especially important in materials science and engineering, has suffered in this atmosphere.

An important function of the federal agencies, and the principal mission of NSF, is to sustain truly innovative research. Projects supported by NSF, especially, should be expected to delve deeply and to take risks, which means that priority must be given to support for innovative individuals rather than large projects.

Research in engineering has become a proper part of the responsibility of NSF, because solving many of today's most important technological problems requires new ideas and a thorough understanding of basic principles. However, in the face of pressing national needs, NSF may be asked to adopt too narrow an interpretation of its mission in engineering—that of searching for short-term solutions by whatever means are available rather than probing deeply to develop fundamentally new technologies. What is needed is a farsighted, interdisciplinary approach to the whole of materials science and engineering.

Collaborative Centers

The question of whether a materials-related center of excellence is beneficial depends on the circumstances involved. The center concept can be most beneficial if it provides a mechanism for several parts of the technical infrastructure to come together, so that center activities amount to more than the sum of its parts. If industrial input is factored into the work, the R&D will have a natural outlet in industrial manufacturing. The training of students and research scientists can often be combined in such an endeavor.

The United States has made three major efforts in center development in the past four decades, and each one has been successful in its own way. The 1940s primarily saw the establishment of major centers (now known as national laboratories) related to nuclear development and high-energy physics. They were established by what was then called the Atomic Energy Commission, and they now belong to DOE. To a large extent, these centers were built around major facilities—for example, nuclear reactors, accelerators, and synchrotron light sources. With them have come major efforts in materials science and engineering to use these facilities for fundamental research and for the development of materials for defense systems, civilian power reactors, and energy systems of many different types. These national laboratories are largely multipurpose, and, increasingly now, they work closely with industry in applied research. The principal point, however, is that the centers were usually built around large and expensive devices that only the federal government could afford, because the projects undertaken were too large and too expensive to be built by industry or universities.

The second group of centers, called the materials research laboratories, was established in the early 1960s in response to the recognition that materials science and engineering was becoming an interdisciplinary field drawing on chemistry, physics, metallurgy, ceramics, and engineering, as well as in response to the belief that the development of many advanced technologies is hampered by the lack of advanced materials. (For a recent history of the origin and progress of these centers, see papers by R. Sproull and L. Schwartz in *Advancing Materials Research*, National Academy Press, Washington, D.C., 1987.) These laboratories, about a dozen in number, were established at major universities by the Defense Advanced Research Projects Agency of DOD, and they brought together different disciplines in an interdisciplinary facility. In 1972, administration of these laboratories was transferred to NSF. Their purpose can perhaps be summarized best by quoting from the government contracts that established the centers within the universities:

The contractor shall establish an interdisciplinary materials research program and shall furnish the necessary personnel and facilities for the conduct of research in the science of materials with the objective of furthering the understanding of the factors which influence the properties of materials and the fundamental relationships which exist between composition and structure and the behavior of materials.

The reasons for the success of these laboratories were summarized by Sproull, and a few are abstracted here from his paper:

The important features at each university were, first, that an umbrella contract provided for continuity of support and for the ability to buy large quanta of equipment and facilities. Second, a local director committed a substantial fraction of his career to making the program succeed. He would use the longevity of support to extract concessions from the university and departmental administrations. Third, the contract provided, in most cases, reimbursement over 10 years for the new construction required to do modern experimentation on materials. Fourth, the longevity of the contract induced the university to allocate the project scarce and prime space in the middle of the campus, thereby establishing the maximum informal connections among disciplines. Fifth, central experimental facilities (such as those for electron microscopy or crystal growth) could have state-of-the-art equipment, even if it was very expensive, and they served as a mixing ground for students and faculty from several disciplines. Sixth, an executive committee composed of people with power and influence in the individual disciplines but oriented toward the success of the program helped the director over the rough spots with department chairmen, people who often were overly protective of their own turf. Seventh, a contract was not given to an institution unless it had a strong disciplinary base on which to build. Eighth, individual grants and contracts with federal agencies continued; most well-established principal investigators received the majority of their support from some other agency and might enjoy help from the program only in the central facilities or the building space (R. Sproull, *Advancing Materials Research*, National Academy Press, Washington, D.C., 1987, p. 31).

The most recent U.S. experience in the development of centers involves the engineering research centers established by NSF beginning in 1985. These centers were formed as a new approach to meeting the serious challenges to U.S. industrial competitiveness. They are located at universities and bring together the capabilities and resources of government, universities, and industry. At cross-disciplinary engineering centers, researchers work on problems that are of technological importance to industry, as contrasted with the work done at the materials research laboratories, which specialize in the generation of scientific knowledge. The centers are formed around specific technological areas, and their titles indicate the specificity of the objective of each center: Systems Research, Intelligent Manufacturing Systems, Robotic Systems in Microelectronics, Composites Manufacturing Science and Engineering, Telecommunications Research, Biotechnology Process Engineering, Advanced Combustion, Engineering Design, Compound Semiconductor Microelectronics, Advanced Technology for Large Structural Systems, and Net Shape Manufacturing. It is planned that several additional centers will be formed in subsequent years.

Paralleling the various NSF centers in many respects, the university research initiatives sponsored by DOD were begun in FY1986. To date, more than 91 initiatives have been funded at universities at a total cost of more than $100 million. The University Research Initiative Program includes fellowships, young investigator awards, exchange scientists, and research instrumentation funding.

The various interdisciplinary research centers supported by federal agencies in collaboration with universities and industry have made valuable contributions to materials science and engineering, and the committee believes that they must continue in the roles that they have served. At the same time, the committee believes that stronger efforts should be made to link science and engineering in interdisciplinary research centers.

FEDERAL LABORATORIES

There are more than 700 small and large federal laboratories supported by 14 federal agencies. Of this total, about 160 federally operated laboratories and about 30 contractor-operated laboratories employ 50 or more scientific and technical personnel.

Based on a survey conducted by the committee, the committee estimates that approximately 2000 people at the national laboratories work in materials science and engineering, divided about equally between the defense and nondefense laboratories.

Department of Energy Laboratories

The national laboratories funded by DOE, with their extensive technical resources and special facilities, are especially well positioned to make major contributions to materials science and engineering, as noted in Chapter 5.

The national laboratories have a long history of involvement in materials synthesis and processing, materials characterization, equipment and instrument development, and systems integration. They also have a history of organizing, operating, and maintaining major collaborative research centers and user facilities for participatory R&D by industries and universities. In general, they have much experience in bringing together multiple disciplines to solve complex problems and in overcoming problems of technology transfer between diverse institutions. However, appropriate incentives need to be developed to help these laboratories reach their full potential for collaboration with industry and universities.

Department of Defense Laboratories

There are 81 DOD laboratories supported through the Navy, Air Force, and Army, with about 90,000 staff members and a total budget of about $1.7 billion—half of the total federal laboratory budget. The property and equipment assets of these DOD laboratories are in excess of $4 billion.

Materials science and engineering is also conducted in about 124 industrial laboratories under the IR&D programs, which are reimbursed by DOD through indirect cost accounts. About 1080 IR&D projects address materials science and engineering issues, and 417 of these have materials science and engineering as their principal subject. The total amount of funding for these projects was approximately $255 million in FY1986.

National Institute of Standards and Technology

The National Institute of Standards and Technology (NIST; formerly the National Bureau of Standards), which receives its funding from DOC rather than DOE, is another potentially valuable resource for materials science and engineering. The approximate budget for materials science and engineering in the NIST in 1986 was $37.8 million, with emphases across the full range of materials classes. The number of full-time equivalent NIST staff members now working in the field is approximately 400; about 200 have doctorates.

Industrial-academic interaction is an important element of the research program at the NIST. The visiting scientist program includes a unique industrial research associate program in which industry sends personnel at its own cost to work collaboratively on joint projects with researchers at the institute, for periods ranging from a few weeks to more than a year. An estimated 20 percent of NIST operating funds comes from industrial sources, mostly via contributions of equipment or the paid time of industrial research associates.

MAJOR NATIONAL FACILITIES

In materials science and engineering, the large-scale national facilities in use today evolved gradually. The first of them, research reactors for neutron scattering, were not even built as materials research facilities but were constructed with funds designated for reactor development. Several synchrotron radiation facilities were originally high-energy physics machines. Thus materials scientists and engineers became accustomed to borrowing time on machines originally dedicated to other purposes.

Beginning in the mid-1970s, the use of synchrotron radiation in materials science and engineering expanded dramatically from an estimated 200 users in the United States in 1976 to close to 2000 users in 1986. The average annual growth rate of published scientific articles based on synchrotron radiation during that period was 30 percent. In coming years, the use of synchrotron radiation facilities will continue to evolve beyond exploration of the frontiers of materials science into more applied efforts. As industrial and governmental use of these facilities matures, an increasing amount of proprietary work will be done at synchrotron radiation facilities. Areas of potential application span a wide range of technologies related to the economy, defense, and health. For example, synchrotron radiation is a potential technology for lithography of semiconductor circuits with submicron feature sizes. It is also associated with the development of the free-electron laser, which has applications in national defense. It has applications in biological imaging. Research on synchrotron radiation will continue to focus on generating increasingly coherent sources of radiation, so that aspects of a laserlike source can be achieved from the ultraviolet to the x-ray region.

The neutron scattering community in the United States has undergone a similarly rapid increase in recent years, having doubled in just 6 years. This upsurge of users from government, industry, and universities is part of a worldwide trend; for example, the neutron scattering community in Western Europe more than tripled during the 1970s. A result will be an increasing demand for use of major facilities.

All of the major national facilities now in operation are open to users. The facility at Brookhaven National Laboratory, for example, still functions as a small science facility, despite the large number of users. Participating research teams, including teams from government laboratories, corporations, and universities, are allowed to organize and carry out their own experiments. One advantage of this arrangement is that industry invests in the instrumentation for research. Another advantage is that many of the universities with materials research laboratories have established beam lines at the facility as part of their funding for the laboratory. This arrangement allows Brookhaven National Laboratory to function in effect as a collection of many different small facilities doing many different types of science.

Thus advanced instrumentation is playing a role of growing importance in materials science and engineering research. This trend justifies the dedication of major facilities to work in the field. A 1984 National Research Council report entitled *Major Facilities for Materials Research and Related Disciplines* (National Academy Press, Washington, D.C.) provides detailed justification for such facilities. That report also reviewed the status of major facilities in the United States and established priorities for future facilities providing instrumentation in synchrotron radiation, neutron scattering, and high magnetic fields. Before laying out its priorities, however, the report highlighted two major recommendations: (1) resources should be provided to operate existing user facilities productively, and (2) the support of major facilities should be accompanied by expanded support of smaller materials research programs.

The priorities established by the National Research Council's Major Materials Facilities Committee were construction of major new facilities and establishment of new capabilities at existing facilities. Assessing only facilities with an initial cost of at least $5 million, the committee recommended construction of (1) a 6- to 7-GeV synchrotron radiation facility, (2) an advanced steady-state neutron facility, (3) a 1- to 2-GeV synchrotron radiation facility, and (4) a high-intensity pulsed neutron facility. With regard to new capabilities at existing facilities, the committee recommended (1) centers for cold neutron research, (2) insertion devices on existing synchrotron radiation facilities, (3) an experimental hall and instrumentation at Los Alamos National Laboratory, (4) upgrading of the National Magnet Laboratory, and (5) enriched pulsed neutron targets.

This committee concurs with the earlier prioritization of major national facilities, and it notes with approval that construction of several of them has recently been started. Current plans call for the construction of the 6-GeV synchrotron radiation facility at Argonne National Laboratory, with the lower-energy machine being built at Lawrence Berkeley Laboratory. According to the most optimistic scenarios, neither of these facilities is expected to be operational before the early 1990s. Part of the experimental floor of the National Synchrotron Light Source is shown in Figure 6.2. Implementation of the highest-priority recommendation for upgrading existing facilities—to build new cold sources, guide halls, and instruments at Brookhaven and NIST—has begun at NIST and when completed will greatly enhance the capabilities of that facility.

The second-priority recommendation in the area of major new facilities and noted aims is for an advanced steady-state neutron facility. This recommendation needs to be fully implemented. There are major opportunities in materials science and engineering if higher fluxes and advanced instruments become available. The increased sensitivity and resolution will greatly extend the range of sizes and level of microstructural features that can be

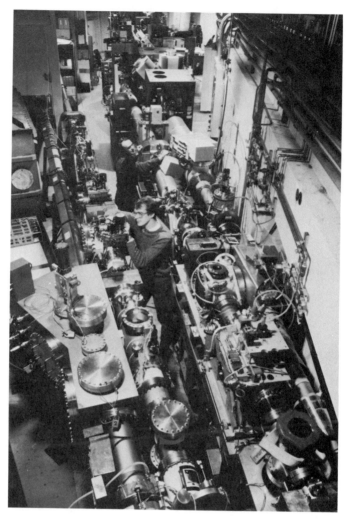

FIGURE 6.2   Shown are two pairs of beams lines coming from the high-energy x-ray storage ring at the National Synchrotron Light Source that are dedicated to materials science. They were constructed by the Naval Research Laboratory and the National Institute of Standards and Technology. On the left pair of lines, x-ray tomography and scattering experiments are performed. On the right pair, soft x-ray spectroscopy and photoemission experiments are performed. (Courtesy Brookhaven National Laboratory.)

studied in bulk materials. It will be possible to follow the kinetics of processes such as precipitation, phase decomposition, coarsening, and damage accumulation in real time. These advanced capabilities for microstructural and nanostructural research and evaluation will provide important fundamental information directly related to the processing, behavior, and reliability of advanced materials.

FINDINGS

Many in the materials field believe deeply that the United States has a critical need for wholly different types of major national facilities, that the country needs facilities that concentrate directly on problems of industrial importance in ways that can help U.S. materials-related industries gain or maintain a competitive edge in the international arena.

A number of state initiatives aim to address this problem. But at the national level the United States has nothing of the scope of the Fraunhofer institutes in West Germany or the quite different Japanese initiatives in the materials field. In 1988 alone, for example, Japan initiated two major new research consortia in the materials field. One of these, in the field of superconductivity, has funding commitments of approximately $26 million per year for 10 years. (This figure does not include the salaries of the estimated 100 researchers to be involved.) The second initiative is in the field of forming of semisolid metals and composites. It is funded at about $30 million for 5 to 7 years (again excluding salaries of researchers).

Modern advanced materials almost invariably require the most sophisticated synthesis and processing facilities. For example, electronic materials such as semiconductor crystals, epitaxial layers, fabricated chips, and optical fibers frequently must meet impurity requirements at the sub-parts-per-billion level to be useful for fundamental studies or applications. Consequently, clean rooms are now a normal requirement for any serious research capability in electronic materials. Typical clean-room costs are in the range of $100,000 to $1 million depending on size and class. Chip fabrication facilities, now a normal requirement for teaching and university device research, are even more expensive. Preparation of many electronic materials requires extremely poisonous reagents. For example, III-V semiconductors often require arsine and metallo-organics that are 10 times more lethal than hydrogen cyanide. A typical laboratory preparative facility for arsine use costs from $100,000 to $1 million.

Molecular beam epitaxy (MBE) gives a finesse in preparation unequaled by any other technique. Control is literally at the atomic level. Novel structures exhibiting new quantum physics can be prepared. Novel devices have been fabricated. Future possibilities, especially those beyond the III-V semiconductors in which most of the work to date has taken place, are just beginning to be realized. However, MBE equipment is not cheap. Again, the range is $100,000 to $1 million for an "entry-level" apparatus. Once an MBE apparatus has been used for a particular class of materials, it cannot be readily converted to the preparation of unrelated materials because of the very high levels of purity required for MBE materials.

Similar expenses are encountered for high-pressure equipment for modern synthesis, for the preparation of modern ceramics, and for polymer processing facilities, among others. Even Czochralski crystal-pulling equipment typically costs more than $1000 per station, and production-size equipment, which is sometimes necessary to study factory conditions realistically, can be 100 times more expensive.

Similar examples related to other materials classes can readily be cited. For example, clean rooms for processing modern ceramics are critical at a number of stages to minimize flaw formation. This and other equipment costs (e.g., for sol-gel and laser synthesis, attrition, and hot isostatic pressing) can readily total in the range of $5 million to $10 million for a sizable laboratory. For metal processing research, rapid solidification equipment, vacuum melting and deformation equipment, and advanced instrumentation for in situ process analysis rapidly bring costs to a similar range.

Understanding of materials at the atomic through the macroscopic levels constitutes the technological base for most modern manufacturing. Only for the most rudimentary operations in manufacturing (e.g., assembly of completed parts), is it possible to control yield, quality, cost, and schedule without understanding, at the most basic level, the materials being processed. This is certainly the case with materials synthesized at the atomic or nanometer scale, but it is also increasingly the case with so-called good, old-fashioned monolithic materials.

Clearly, modern synthesis and processing are neither string and sealing wax nor beaker and Ehrlenmeyer flask activities. Synthesis and processing require substantial investment in equipment. Industrial laboratories usually realize these needs and generally support synthesis and processing activities at an appropriate level. The understanding that synthesis and processing require large expenditures in academia and government laboratories must be more widespread if materials science is to have the steady stream of trained people, new materials, and well-controlled processes that are essential for realization of its potential.

The committee, therefore, recommends substantially increased expenditures on facilities for synthesis and processing research in universities and other laboratories—for use by individual investigators and small groups of investigators. It also recommends establishment of major national centers for synthesis and processing, involving universities, industry, and government. Such centers could be expected to focus on generic synthesis and processing, or on a given materials class. Some would also be expected to have an advanced instrumentation component (e.g., synchrotron light sources for semiconductor processing). Some would be expected to involve close coordination with disciplines outside the materials or materials-related fields (e.g., business). All would focus on development of industrially useful technology, technology transfer, and quality and economy in design and production processes.

7

Comparisons of Efforts in Materials Science and Engineering of Selected Nations

The President's Commission on Industrial Competitiveness noted in its 1985 report *Global Competition: The New Reality* (U.S. Government Printing Office, Washington, D.C., 1985), "For this entire century—until 1971—this Nation ran a positive balance of trade. Today, our merchandise trade deficit is at record levels." There are many reasons for this decline, including economic, political, and social forces well beyond the scope of this report. But as previous chapters have demonstrated, materials science and engineering is a key determinant of manufacturing success, and efforts in the field will be crucial to recovering and retaining the U.S. competitive edge.

To assess international cooperation and competition in materials science and engineering, one of the committee's panels gathered information through a questionnaire sent to materials science and engineering leaders in competitor countries and obtained additional data from the science attachés of foreign embassies, from case studies on representative industrial sectors, and from the open literature. The panel then compared the activities of other countries with practices in the United States. The countries surveyed included several traditional U.S. trading partners (Canada, West Germany, France, and the United Kingdom), a major economic competitor and strategic ally (Japan), a newly industrialized country (South Korea), and the principal U.S. strategic competitor (the Soviet Union).

The most striking observation supported by the gathered information is that all the major nations are strongly committed to industrial growth stimulated by coordinated R&D in which materials science and engineering is a featured element. In fact, of all the industrial areas in which growth is anticipated for the next decade, materials science and engineering, biotechnology, and computer and information technology were targeted by all of the nations surveyed.

Another significant observation is that cooperative mechanisms, fostered by government involvement, are being used increasingly by competitor nations to enhance industrial competitiveness. As demonstrated by the industry surveys in Chapter 2, materials science and engineering is rarely the driving force in industrial advancement, except in materials-producing industries, but it is crucial in areas of changing technologies. The complexity of modern manufacturing has led inevitably to interdependence among industries. This trend is on the upswing, taking the form of joint ventures, licensing and use of outside sources for manufacturing, and cooperation in the long-term R&D of technologies for improved manufacturing capability. Such cooperation is most advanced in Japan, where it is mediated by government funding and is often carried out in government laboratories in collaboration with industry. In the United States, cooperation among industries is accomplished through industry-sponsored research-granting organizations, R&D laboratories sponsored by industrial consortia, and various industry-university centers. Noticeably lacking in the United States, and found to a greater degree in all of the countries studied, is a national agency charged with stimulating and assisting industry and, when appropriate, with ensuring that cooperative activities are coordinated and their impact on industrial development optimized.

Other important conclusions derived from the analysis of international competition and cooperation in materials science and engineering include the following:

• The views of industry, universities, and government are sought and received by the governments of foreign countries. In the United States, however, this input is informal. Most other nations set directions for materials science and engineering in a manner intended to target specific industrial markets. In the United States, there is no official materials science and engineering strategy.

• Foreign governments universally try to ensure the coupling of R&D with commercial exploitation of research results. The use of government laboratories to achieve this is common to most other nations, with the general lack of such activity in the United States a significant difference.

• The availability of adequate numbers of trained materials scientists and engineers is a concern of all nations, but control of the educational system varies greatly among the countries surveyed. The extremes on the spectrum of control are represented by the United States, with its vast decentralized system of higher education, and South Korea, where levels of educational funding are tied directly to the gross national product and technical training areas are emphasized as part of the national economic plan. All of the countries surveyed indicated that their educational emphasis in materials science and engineering had increased during the period 1976 to 1986 relative to other areas, with further emphasis expected in the next 10 years.

• As in the United States, materials science and engineering is taught academically in a variety of departmental settings in all of the nations surveyed. In all of the countries except Japan and South Korea, the trend is toward materials science and engineering becoming increasingly multidisciplinary. Research emphasis in academic departments is similar throughout the world: 30 to 50 percent is applied research, and the rest is basic research. South Korea is a striking exception, with 80 percent of the university research described as applied. There seems to be a general trend toward conducting more applied research at universities, although this is not the case in Japan (where university-industry links are traditionally not close), South Korea (where the links could hardly be closer), and West Germany (where the more applied work is carried out in the Fraunhofer laboratories, which are only loosely tied to the universities).

• Government policy and funding for materials science and engineering education are viewed as marginal to only moderate. Inattention to and lack of funding for education in materials science and engineering appear to be major oversights in all of the nations surveyed.

• Techniques for implementing national goals for materials science and engineering are similar throughout the world, with centralized planning and implementation, establishment of science and technology programs with definite objectives, and cooperative mechanisms being the favored tools of many countries.

MATERIALS SCIENCE AND ENGINEERING ABROAD

Canada

Canada ranks seventh in the world in gross national product and sixth in trade after the United States, West Germany, the United Kingdom, France, and Japan. Since 1984 Canada has achieved one of the highest economic growth and job creation rates among the Western nations, due in part to its spectacular growth in manufacturing. Industry is now a leading component of the nation's economy and employs about one-third of its work force. Government concentrates its research on areas to improve its industrial trading position through support of basic research, in tandem with providing inducements (e.g., R&D tax credits and capital gains exemptions) to industry for developmental efforts.

The Canadian view is that their traditional industries and markets no longer can be counted on to fully sustain the nation's economic growth, and new advanced technologies must be fostered through cooperative R&D efforts. A comprehensive federal science and technology policy has been under development that focuses on strategies for increased R&D expenditures by the private sector to complement federal and provincial initiatives. The Ministry of State for Science and Technology, the government organization most responsible for science policy and coordination, has identified three areas as strategically important to the country's future: information technology, biotechnology, and advanced industrial materials. Current federally sponsored R&D provides a key network of activities; in the area of advanced industrial materials, government R&D amounted to about Can$29.7 million for the period 1985 to 1986.

The organizational setting for these R&D programs is pluralistic and coordinated and is built around the federal-state system of government in which policy and programs must be in accord between the federal and provincial governments. Line departments—such as the Department of Regional Industrial Expansion, the Department of Energy, and the Department of Mines and Resources—and the Canadian National Research Council and the Natural Sciences and Engineering Research Council (NSERC) have the ultimate charter for implementing federal science and technology policy. Each manages the part of Canada's research budget (about Can$2.9 billion in 1982–1983) within its own jurisdiction. The Canadian National Research Council is among the top funding agencies for R&D (about Can$361 million in 1982–1983); its charter mandates industrial expansion and regional development. The Canadian National Research Council operates its own laboratories, gives direct financial support to universities and industry for specific R&D projects, and sponsors coordinating research activities. The NSERC underwrites university research (Can$227 million in 1982–1983), supplementing provincial funding. Materials science and engineering projects focus primarily on metals, but studies of other materials such as advanced ceramics, composites, and polymers are increasing.

West Germany

The distinctive feature of West German industrial and economic policy, planning, and programs is their broad-based consensus-building process. This democratic process combines elements of decentralized decision making and regional implementation, with sectoral autonomy a key concept and representation by major interest groups a guiding premise. These characteristics also typify the science and technology system in West Germany and set it apart from the approaches used by other European countries. Materials science and engineering has received long-standing emphasis in many West German R&D programs. The lead organization for science and technology policy is the Ministry of Research and Technology (BFMT). The BFMT receives about 60 percent of all federal R&D funds (about DM 7 billion in 1984). About half of these funds go directly to industry on a 50:50 cost-shared basis. The remainder is used to support major national research centers, educational institutions, and private research organizations, many of which have an industrial focus.

A significant fraction of research funding is channeled to a series of quasi-independent research institutes or laboratories through major nongovernmental research associations or societies, including the Max Planck Society (basic research), the Fraunhofer Society for Applied Research (industrial research), and the German Research Society (education). The research institutes of these societies are usually small, highly focused, and autocratically administered. The Fraunhofer Society for Applied Research, for example, consists of 34 separate institutes and employs about 4000 people, one-third of whom are scientists and engineers. Its institutes cover nine important industrial areas: microelectronics, information technology, automation, production technologies, materials and component behavior, process engineering, power and construction engineering, environmental research, and technological economic studies and technical information. The materials and component behavior area ranks first in terms of staff allocation (with approximately 500 employees) and second in budget (behind microelectronics, each with about DM 53 million in 1985).

In 1985, the BFMT inaugurated a new 10-year materials research program with an annual budget of about DM 100 million. The BFMT has assigned the Nuclear Science Research Center at Julich to manage the new program, which encompasses the following areas: high-performance structural ceramics, powder metallurgy, high-temperature metals and special materials, high-performance polymers, and advanced composites. About 30 institutes, representing the Fraunhofer Society for Applied Research, the Max Planck Society, and West Germany's large research centers, cooperatively participate with numerous industrial companies in this program.

France

France has developed a modern and highly diversified industrial enterprise that generates about one-third of its gross national product and employs about one-third of its work force. It is now a major exporter of steel, chemicals, motor vehicles, nuclear power, aircraft, electronics, telecommunication products, and weapons, with the latter five targeted by government for industrial advancement. National planning and policymaking in France are unified for all areas, including R&D. They are highly centralized within a governmental system structured for maximum coordination and control of programs.

Organizationally and operationally, the R&D system in France is enmeshed in an interministerial structure, each ministry covering different spheres such as defense, industry, and education. The Ministry of Research and Technology focuses government R&D on national industrial technology programs, as well as providing oversight and management of the nationalized industries. Within the ministerial system, the government operates a host of research establishments and laboratories. By far the most extensive and important agency for R&D is the Centre National de la Recherche Scientifique (CNRS). Attached to the Ministry of Education, it is organized along the lines of the traditional academic disciplines and supports primarily basic research. CNRS does not have a research directorate for materials science and engineering, but it has established crosscutting programs in communications science and new materials. In 1988, CNRS had a budget of about Fr 9.0 billion, about 24 percent of the total civilian R&D expenditures, and employed almost 10,000 scientists and 15,000 support staff in 1,350 laboratories or universities, other government agencies, and industry.

As a complement to their internal research efforts, the French have sought to extend their technology base through international cooperative programs. For the most part these are geared toward industrial development and involve multination participation under the auspices of the European Communities. The two most notable programs are the European Strategic Program for Research in Information Technology (ESPRIT) and the European Research Coordinating Agency (EUREKA). The latter is the French equivalent of the U.S. Strategic Defense Initiative program but is oriented to technology, not defense. Both programs require industry participation in funding and conducting research.

The United Kingdom

Research and development in the United Kingdom is extremely pluralistic and decentralized and in many respects resembles the U.S. R&D system in that policy and planning are carried out by several government departments. Although new programs have been established in the United Kingdom and new approaches such as collaborative research are being tried, the elemental organization of R&D has remained fairly static over the years. On the whole, no group within government coordinates its $6.1 billion per year R&D program, and individual departments operate autonomously. Research and Development (ACARD) is the main body influencing coordination of applied R&D between government and external groups. However, it has no management function, nor does it allocate resources; it does provide the primary conduit for industry access to top government department heads.

Most university research funds come from the government's budget and are administered by the Department of Education and Science. In 1983 the department spent about $1 billion on university research, a sum that included major funding for four major research laboratories operated by the Advisory

Board for the Research Councils. Defense R&D currently consumes more than 50 percent of the United Kingdom's research budget. The Ministry of Defense provides this support primarily to industry via contracts and for operation of its own set of laboratories. Less than 2 percent of the ministry's budget is used for basic research at the universities.

The principal government agencies for civilian R&D are the Department of Trade and Industry and the Department of Education and Science, with some added activity by the Department of Energy. The Department of Trade and Industry supports industry in two ways: (1) by direct investment (e.g., loans and preproduction guarantees) in firms through its National Research Development Corporation and (2) by direct R&D contracts, usually on a cost-shared basis. In 1983, 61 percent of its funds were spent this way; the balance of the department's resources went to support programs in other government departments and in its own laboratories. Today, there is a general redirection of the United Kingdom's national research establishments to R&D more related to market-oriented needs. Research organizations such as the National Engineering Laboratory, the National Physical Laboratory, and Harwell Laboratory work with industry on a contract basis. Harwell, for example, operates essentially as an independent laboratory, serving industry in a self-sufficient fiscal mode.

On the whole, industry contributes less of its own money to R&D than the government spends, a practice just the opposite of that in most other Western nations. British industry is a mixture of public and private firms, and to aid it, the major new 5-year, $500 million Alvey program was established in 1983, primarily to bolster the United Kingdom's competitive position in microelectronics. The program follows a consortium model involving cooperative R&D, with the costs shared between industry and government on about a 50:50 basis. A follow-on Alvey program ($1.58 billion) is now under consideration, and initiation of still another major collaborative program aimed at developing high-technology products is anticipated. The $640 million Link program will make funding available for selected university projects, provided that the costs are shared equally with industrial sponsors. Projects will cover molecular electronics, transportation systems, food processing engineering, and materials technology. It is presumed that the basis for the projected R&D on materials technology under the Link program had its origin with the submission in 1985 to the Department of Trade and Industry of the Collyear report. The Collyear Committee proposed a 5-year, £120 million program for the wider application of new and improved materials and processes.

Japan

The Japanese materials science and engineering establishment is a highly structured enterprise that has been instrumental in many past technological successes. However, it is composed of conventional organizational elements and implementation and strategy instruments quite similar to those used throughout the world. What is atypical in Japan is its systems approach— its long-term and consistent policy, stimulated and coordinated by government but coupled to an effective communication link between the public and private sectors, including a multilevel advisory committee arrangement. In an orchestrated division of activities and responsibilities, the government acts as the catalyst, and industry takes the lead role as a funder and performer of R&D.

Within government, the highest policymaking body for R&D is the Office of the Prime Minister. Two advisory councils, the Science and Technology Council and the Science Council, provide guidance on technology and on pure science matters. The members of these councils are leading spokespersons within and outside of government; the chairman is the Prime Minister of Japan. These councils establish national goals and provide broad directions for science and technology in general and materials science and engineering in particular. They have a great impact on Japan's yearly federal R&D budget, which was about ¥1500 billion in FY1984.

The Ministry of Trade and Industry (MITI), the Ministry of Education, Science, and Culture, and the Science and Technology Agency (STA) essentially share government operational responsibilities for materials science and engineering, including planning, funding, and oversight.

The Science and Technology Agency is located within the Prime Minister's office. It receives about 27 percent of government R&D funds for major national projects such as the space and the reactor programs. The agency also has a mandate to stimulate basic research within industry and, through its Japan Research Development Corporation, to support new technology developments, such as the Exploratory Research for Advanced Technology, using start-up companies as one mechanism of implementation. Attached to STA are six research institutes, two of which, the National Institute for Research in Inorganic Materials and the National Research Institute for Metals, are the principal laboratories most related to materials science and engineering. Under STA, they often perform R&D in cooperation with MITI, the industrially oriented ministry.

The Ministry of Education, Science, and Culture administers about 47 percent of government research funds, all of which it provides to the universities and national centers for scientific research.

The Ministry of Trade and Industry is the central government organization with industrial development as its primary charter. It receives only about 13 percent of government R&D funds and relies on cooperative mechanisms with industry to leverage considerably more R&D. MITI formulates industrial technology plans, determines and provides for subsidies and/or funding, and selects participating industrial R&D groups and associations to work with MITI's 16 national laboratories. The national laboratories fall under the jurisdiction of one of MITI's operational arms, the Agency of Industrial Technology and Science (AIST), which in FY1985 had a budget of ¥122 billion. A sister agency, the Japan Industrial Technology Association, functions as the licensing agency of AIST and provides regular information on foreign technology developments.

Typical of MITI's procedural mode is its program on advanced materials, the R&D Project on Basic Technology for Future Industries. This program, under the auspices of AIST, targets three general research areas—biotechnology, electronics, and advanced materials. Generally, AIST forms a nongovernmental advisory committee for each project, and an industrial association is created to work cooperatively with all other members of the organization and of MITI's national laboratories.

To complement Japan's already complex industry-government cooperative agenda, a new dimension has recently been added. In October 1985, the Diet established the Japan Key Technology Center to be run under the joint oversight of MITI and the Minister of Posts and Telecommunications. The Key-TEC program (estimated to cost about ¥31 billion) is viewed as part of a needed effort to bolster science by supporting long-range applied and fundamental research on key, but very new, advanced technologies. The focus of the program is to be about 10 years out in front of current knowledge and is supposed to result not so much in prototype products as in generic information on which later development of products can be based. Because of the advanced technology mission of Key-TEC, the program can be described as a Japanese civilian analog of the U.S. Department of Defense's Defense Advanced Research Projects Agency.

South Korea

The rapid industrial development of South Korea matches or even surpasses that of Japan, and for many of the same reasons. Industrial developments proceed rapidly because of a strong government that places science and technology in a favored position and rewards the corporations and organizations that are most successful in promoting international trade.

The Korean Advanced Institute of Science and Technology is the largest government supporter of materials science and engineering in South Korea. Overall, materials research in South Korea is divided into two major categories: (1) conventional materials (improvement and import reduction) and (2) technology development (advanced materials). The former category is supported primarily by industry, whereas research in the latter category is financed almost exclusively by the government in a public-private cooperative system. In 1985, there were about 29 advanced materials projects under way in South Korea that included efforts in metals, polymers, composites, and fine ceramics.

There are about 3000 Ph.D.'s working in science and technology in South Korea, with about 10 percent of those involved in materials science and engineering.

The Soviet Union

Science and technology in the Soviet Union is intimately linked with the machinery of government. The Soviet science and technology effort is the most highly structured, centrally controlled system in the world. Five-year plans are formulated by the State Planning Committee (Gosplan) through a coordinated process involving the Academy of Sciences, the State Committee for Science and Technology, and the various other ministries. Within this organizational complex, the Academy of Sciences carries the most influence. Today, the academy is the scientific side of Soviet science and technology, and the ministries are the technological side. Higher science education is handled by both the academy and the Ministry of Higher and Secondary Education. The academy and other educational institutions, as well as all the ministries, operate an array of research establishments of varying size and sophistication, involving well over 1 million workers. Soviet science on the whole is highly rated and in some cases produces enviable results to be watched and built on, as, for example, the Japanese have done in advancing the published Soviet materials and processing developments in the areas of low-temperature diamond film deposition and electrodeposition of fibers for metal matrix composites. Soviet product design and manufacturing technology are inefficient and, more often than not, are characterized by reverse engineering of Western-made goods, a practice leading to a 5- to 10-year lag in Soviet marketing of products.

There is no official tie between any major research groups; thus many of the basic, innovative ideas (including those in materials) generated by the academy's research institutes are not developed in the Soviet Union because the ministries conduct about 90 percent of all engineering R&D and generally do not take an interest in academy business (and vice versa). Although there is superficial coordination, there is no incentive for collaboration, and Soviet industry opts for adaption of Western technology rather than developing its own. As a consequence of this division, materials science and engineering is treated as materials science on the one hand and as materials engineering on the other; the former is generally excellent, and the latter, duplicative.

MATERIALS SCIENCE AND ENGINEERING IN THE UNITED STATES

In the United States, government-sponsored materials R&D, because of its multifunctional and widespread impact, is diffused throughout a multitude of government programs. No single upper-level agency in the United States has the sole mandate for materials or materials science and engineering. As described in Chapter 2, materials R&D pervades the activities of the major government agencies, but always within the context of their specific missions. As a result, government-related materials science and engineering, and science and technology in general, are pluralistic and decentralized in the United States.

The organizational framework for materials science and engineering in the United States is similar to that for other nations. However, unlike many of its competitors, the United States does not have a major agency charged with fostering industrial advancement and with coordinating and integrating the spectrum of materials R&D activities on which industry depends. In Japan, for instance, MITI is a strong force in industrial affairs and materials science and engineering; nothing comparable exists in the United States. Accordingly, the direction of materials R&D taken by the U.S. government is the sum of all the directions of the parts making up the R&D system. Coordination and control vary from agency to agency, and national priorities emerge from the perceptions of national needs and opportunities held by individual agencies, guided by cabinet-level policies and directions.

For its overall planning, the U.S. government relies on formal and informal advisory groups and organizations at all levels within and outside government. For the most part, however, industry and universities provide science and technology policy advice only through informal communication links. Although many separate agencies have statutory advisory groups and Congress hears testimony from individuals and groups, there are no standing national councils involving industry-university-government participants for joint planning, coordination, and program evaluation. Dialog between the public and private sector, and within sectors generally, is not organized and does not occur on the scale found in competitor nations.

During the 1970s the Office of Science and Technology Policy (OSTP) and the Office of Technology Assessment were created to advise the President and Congress, respectively, on R&D issues as a whole, including materials issues. In 1982 a science council reporting to OSTP was established to improve coordination of the national research effort. OSTP also chairs a coordinating Committee on Materials (COMAT), made up of representatives of government agencies engaged in materials R&D. The Office of Management and Budget provides further oversight through its budget review and approval process. The General Accounting Office, an analytical arm of Congress, furnishes additional assessments and advice. Major independent sources of advice to the government include the National Research Council.

Important legislation affecting materials science and engineering includes the 1980 National Materials and Minerals Policy, Research and Development Act, which required coordination by the President of the government's minerals and materials activities. This was followed by the National Critical Materials Act of 1984, which called for the establishment of a National Critical Materials Council, the establishment of a national federal program for advanced materials research and technology, and the stimulation of innovation and technology application in the basic and advanced materials industries. As of this writing, implementation of the law by the executive branch is still in the early stages. Congress also enacted the Cooperative Research Act of 1984, which, along with proposed legislation modifying the Clayton Act, provides a more favorable environment for cooperative R&D between businesses, in part by reducing antitrust penalties.

Government provides slightly more than one-half of the more than $120 billion currently devoted to all types of R&D in the United States; industry provides the balance. Definitive statistics are not available on industrial funding of R&D in materials science and engineering, but industrial funding is believed to be greater than the $1 billion spent by government. Government-sponsored R&D is carried out by contract in industrial, university, and independent laboratories and in university research centers; it is also carried out with funds from direct congressional appropriations in the government's own departmental laboratories, in federally funded R&D laboratories—principally the national laboratories—and in research centers sponsored by the National Science Foundation (NSF). Most federal R&D funds go to defense-related research. For materials-related R&D, the Department of Energy sponsors more than 50 percent of the work; the balance of the funding is provided principally by the Department of Defense (DOD), the National Aeronautics and Space Administration (NASA), and NSF, with smaller efforts funded through the Department of Commerce [National Institute of Standards and Technology (NIST), formerly the National Bureau of Standards] and the Department of the Interior (Bureau of Mines). The specific research programs on materials are diverse, and they cover most of the materials classes and types, but usually in the context of broad efforts such as engine or very large scale integration (VLSI) development. Industry performs about 75 percent of all the R&D conducted in the United States. It spends most of its own R&D funds in its own laboratories and the rest at independent research centers and the universities. Corporate R&D expenditures are often reported and analyzed as a percentage of sales, so that R&D, particularly long-term R&D, may suffer from the vagaries of the near-term economic climate.

MECHANISMS FOR COOPERATIVE RESEARCH

Cooperative research entails the joining of technical and financial resources to pursue areas of collective interest to achieve specific individual goals. Recent times have seen the methodical creation and buildup of a plethora of new technical linkages among businesses and research organizations throughout the world, outstripping past efforts. These take many forms, and joint ventures, multinational corporations, national and international consortia, and an array of new types of collective industrial research associations now abound. However, both the concept and conduct of cooperative R&D involving private corporations are more common in Europe and Japan than in the United States. This difference derives partly from the smaller domestic or regional markets, and hence the smaller resources for R&D, in other countries and partly from distinct philosophical convictions regarding competitive behavior. Whatever the reasons, cooperative industrial R&D plays a more active and pivotal role in national affairs overseas than it does in the United States. Increasing numbers of nations rely heavily on government-orchestrated technology development programs in which collaborative arrangements between government, universities, and industry are integral to their strategic approach.

In many European countries there is an extensive network of industry-specific collective associations with independent laboratory facilities, usually operating with a government subsidy along with some formal basis for industry funding. In addition to these strictly national efforts, R&D conducted under the auspices of the European Economic Communities (EEC) represents one of the most extensive collaborative efforts in existence. It involves more than 1 million workers, major research laboratory centers, and a multibillion dollar budget. Recent EEC programs have focused on cooperative R&D requiring direct participation and funding by private firms. Examples of programs relevant to materials science and engineering include ESPRIT, Basic Research in Industrial Technologies for Europe (BRITE), and European Research in Advanced Materials (EURAM).

Japan probably has the most prolific system of cooperative research programs and organizations. Major categories consist of at least 18 government centers, 600 local centers, and many semipublic groupings. The major industry-specific cooperative R&D efforts are funded primarily by MITI and are conducted through more than 50 research associations as authorized by Japan's Industrial Technology Law. Advanced materials for future industries are a featured item on MITI's collaborative R&D agenda.

A distinct feature of U.S. cooperative R&D activities is their diversity. There is no cohesive approach to R&D. Individual researchers, universities, private corporations, and all levels of government participate to different degrees and at different times to meet specific but individual needs. Although the United States has no direct cooperative system, organizational framework, or national policy comparable to those of its competitors, some marginal improvement in this direction is evident. Antitrust laws have been modified, and industrial consortia (e.g., Microelectronic and Computer Corporation, Semiconductor Research Corporation, and the new Sematech) are on the rise. Executive orders are in place to promote better use of the national laboratories by industry. New NSF-sponsored engineering research centers are being set up. State-initiated technology incubator programs are appearing with greater regularity. Still lacking, however, are government-sponsored national laboratories for applied industrial research, which have been seen to be so effective in Japan with its MITI laboratories and in West Germany at the Fraunhofer institutes.

COMPARATIVE ANALYSIS OF U.S. COMPETITIVE STATUS IN MATERIALS SCIENCE AND ENGINEERING

Table 7.1 compares the competitive status of materials science and engineering in the United States with that of materials science and engineering in other countries. It represents the views, as of early 1987, of a group of materials scientists, engineers, and industrialists familiar with the state of the field in the United States and abroad. The assessment was prepared by a Committee on Materials Science and Engineering panel and reviewed by materials science and engineering experts at NIST. The analysis presented here correlated favorably with a similar independent study conducted by NIST. The countries evaluated (Japan, France, the United Kingdom, West Germany, and South Korea) are representative trading partners of the United States viewed as having, or likely to have in the longer term, a competitive advantage in world markets influenced by materials science and engineering. China and the Soviet Union are not included because materials science and engineering operates under different political systems in those countries. Materials science and engineering in Canada and in some newly industrialized countries is typical of that carried out in nations covered in Table 7.1 (France, West Germany, Japan, South Korea, and the United Kingdom) and therefore is not separately assessed. The analysis is intended to illustrate different materials science and engineering systems and the efficacy with which each system works to enable a country to achieve a particular competitive status. Since general science and technology and materials science and engineering are closely related, a major fraction of the comparison is based on science and technology comparisons.

The analysis in Table 7.1 has been grouped under three somewhat arbitrary headings that influence the U.S. competitive position in materials: (1) industry factors, (2) technology factors, and (3) government factors.

Industry Factors

Industry factors have been divided into seven categories: (1) comparative advantage in major markets, (2) comparative advantage in materials science and engineering, (3) productivity, (4) industry structure, (5) innovation to commercialization capacity, (6) resource factors, and (7) capital and financial

TABLE 7.1   Comparative Analysis of U.S. Versus Foreign Competitive Positions (of France, West Germany, Japan, South Korea, and the United Kingdom) in Materials Science and Engineering in 1987

Factors	Trend in U.S. Competitive Position[a]	Current U.S. Competitive Position[b]				
		Disadvantageous			Advantageous	
		Major	Minor	At Parity	Minor	Major
Industry Factors						
Comparative advantage in major markets						
Aerospace	dec.				F	G,J,K
Motor vehicles	dec.	J		G,F		K,UK
Electrical/electronic	dec.		J			F,G,UK
Instruments	n.c.				G,J	F,K,UK
Machinery	dec.		J	F,G,UK	K	
Chemical/allied products	dec.		K	F,G,J		UK
Comparative advantage in materials science and engineering						
Metals	dec.			G,J,UK	F,K	
Ceramics	dec.		J	G,UK	F	
Polymers	n.c.			G,UK		F,J
Composites	imp.			F,G,J		K,UK
Productivity						
Current	dec.			G,J	UK	F,K
Growth rate	dec.	J,K		F,G		UK
Industry structure						
Integrated	imp.			F,G,J	UK	K
Size (big, small niche)	n.c.			F,G,J	UK	K
Multinational	n.c.			F,G,J		K,UK
Innovation to commercialization capacity	dec.	J		G	F,K	UK
Resource factors						
Labor costs	dec.	K	J	F,G,UK		
Labor quality	dec.		G,J	F,K	UK	
Capital and financial factors						
Capital costs	dec.	J,K		F,G,UK		
Long-term R&D investments	dec.	J,K	F,UK	G		
Financial/banking environment	n.c.	J,K	F	G,UK		

Continued

TABLE 7.1 Continued

Factors	Trend in U.S. Competitive Position[a]	Disadvantageous Major	Disadvantageous Minor	At Parity	Advantageous Minor	Advantageous Major
Technology Factors						
Materials sciences and engineering R&D emphasis by task						
Basic	n.c.			G	F,J,UK	K
Applied	dec.		J	F,G	UK	K
Developmental	dec.	J		F,G	UK	K
Manufacturing	dec.	J	K	F,G	UK	
R&D emphasis by material						
Metals	dec.			G,J	F,UK	K
Ceramics	n.c.		J	G,UK	F	K
Polymers	dec.			G,UK		F,J,K
Composites	imp.			F,G,J		K,UK
Materials science and engineering resources						
Funds						
Metallurgy	dec.			UK	F,G,J	K
Ceramics	imp.		J		F,G,UK	K
Polymers	n.c.		G	UK	F	J,K
Composites	imp.		F	G,J		UK,K
Electronic materials	dec.		J		F,G,UK	K
Optical materials	n.c.			J	F,G,UK	K
Manpower education						
Materials science and engineering	imp.			J	F,G,UK	K
Metallurgy	dec.			G	J,UK	F,K
Ceramics	imp.		J	F,UK	G	K
Polymers	n.c.		G		F,J,UK	K
Composites	imp.		F	UK	G	K
Electronic materials	imp.			G,J	F,UK	K
Optical materials	imp.			G,J	F,UK	K
Facilities and equipment						
National/user	n.c.			F,K,UK	J	K
Regional/local	dec.	J	G	F,K,UK		
Funds						
Government (nondefense)	dec.	J,K		G	F,UK	
Industrial investment	n.c.	J,K		F,G	UK	
Interactions and interfaces						
Cooperation (science)						
In-place mechanisms	n.c.		F,G,J,UK			K
Mechanism utilization	n.c.		F,G,UK	UK		K
Cooperation (technology)						
In-place mechanisms	imp.	G,J	F,K	UK		
Mechanism utilization	imp.	G,J	F,K	UK		

Continued

TABLE 7.1   Continued

Factors	Trend in U.S. Competitive Position[a]	Current U.S. Competitive Position[b]				
		Disadvantageous		At Parity	Advantageous	
		Major	Minor		Minor	Major
Government Factors						
National industrial policy						
Structure/organization	dec.	J,K	F	G	UK	
Government-industrial relations	dec.	G,J,K,UK		UK		
Strategy	dec.	J,K	F	G		
Industrial development						
Financial incentives	dec.	F,J,K	G	UK		
Defense-related barriers	dec.	G,J,K	F	UK		
Government services (patents, information, statistics)	n.c.			F,G,J,UK		K
National factors						
Exchange rate	imp.		G,J,UK	F,K		
National debt	dec.	J	F,G,UK		K	
Employment	n.c.	J	K,UK	F,G		
Inflation	imp.			F,G,J,K,UK		
Trade policy	n.c.	J,K	F,G	UK		
Competitive attitude/national prestige	n.c.	J,K	G	F		UK

[a]Trends are as follows: "dec." is declining, "imp." is improving, and "n.c." is no change.
[b]Countries are abbreviated as follows: F is France, G is West Germany, J is Japan, K is South Korea, and UK is United Kingdom.

factors. In the first five of these categories, with a few exceptions (primarily involving Japan), the United States is seen either to be at parity with or to have a clear current advantage over the five countries compared. It is the perception of the experts, however, that in all but 2 of the 16 subcategories, the U.S. position is static or deteriorating.

In categories 6 and 7, by contrast, the United States is seen as having an advantage in only one of five subcategories and as having a static or declining position in all of them. These perceptions are clearly related to some of the subcategories in the government factors section of Table 7.1. They are also important in themselves as guides to government policymakers and to technologists struggling with strategies that seek to reverse other declining trends through new investment in capital equipment. A major question is whether the needed capital will be available.

Technology Factors

The situation described by technology factors is more complex and includes four categories: (1) materials science and engineering R&D emphasis by task, (2) R&D emphasis by material, (3) materials science and engineering resources, and (4) interactions and interfaces. In the first category, the United States is seen to have a clear advantage (and to be holding it) in the area of basic research. In applications, development, and manufacturing—areas that presage the development of new products and processes—the United States is less well off. Japan leads in each of these areas, and the relative position of the United States in each is deteriorating. This situation is clearly related to the poor competitive position of the United States in the area of government-business relations. The United States has a long tradition of government support for basic research but essentially no tradition of direct support for nondefense industrial technology.

The conclusions to be drawn about the second category, R&D emphasis by material, are less clear. The United States has advantages in each materials area and a clearly improving position in composites. In the other areas, despite apparently declining U.S. positions, the headings (metals, ceramics, and polymers) are too broad to reflect some of the more focused strategies within industry and government (e.g., rapid solidification, low-temperature cements, and electronic polymers). Nonetheless, perceptions of deteriorating U.S. leadership in these areas suggest a need for continued monitoring.

The same comments apply to the third category, materials science and engineering resources, which include education, facilities and equipment, and funds. Declining or improving positions cannot be rated as either bad or good in the absolute. To be meaningful, they have to be compared to what would be appropriate under a U.S. materials strategy, and in this category there appears to be none.

Government Factors

Government factors have been divided into three categories: (1) national industrial policy, (2) industrial development, and (3) national factors. Under the first category, national industrial policy, all three subcategories (structure/organization, government-industry relations, and strategy) show the United States in an increasingly disadvantageous position with respect to its competitors. The first two of these subcategories contribute to the third. The United States appears to have neither the structure nor the government-industry relationships that can lead to a national materials strategy that is respected by both business and government.

The absence of an established U.S. materials science and engineering strategy raises fundamental questions. For instance, is the declining U.S. position in technology (e.g., in the steel industry) appropriate to a country at the stage of development that has been reached by the United States, or is it a result of a lack of strategic thinking?

Although Table 7.1 does not represent a statistical survey, it has major significance as the combined perception of materials experts. These perceptions contribute to decisions made in the field, even though they remain perceptions. The overall picture conveyed by Table 7.1 is that of an aging nation that has yet to develop the strategies, structures, or mechanisms to defend its decaying leadership in the world of materials. The "Summary, Conclusions, and Recommendations" chapter suggests ways in which materials science and engineering can be strengthened and applied to areas of national importance.

FINDINGS

For the past 40 years, the United States has been the world industrial leader because of its dominant position in science and technology. During the past decade this position has been deteriorating rapidly as Western Europe and Japan have assumed an aggressive role in technology development, both for domestic and for export markets. In many areas, including materials, these nations now are fully competitive, and in some cases, they have surpassed the United States. Their reemergence in materials science and engineering benefits the field as a whole, but the United States can and must regain its competitive edge. Without it, an essential factor in maintaining U.S. economic well-being will be lost. Foremost among the observations discussed in this study is the strong commitment to industrial growth by all major competitor nations, stimulated by coordinated R&D in which materials science and engineering is a featured element. Indeed, of all the industrial areas in which growth is anticipated for the next decade, materials science and engineering ranks in importance with biotechnology and computer and information technology as the areas targeted for development by all nations sampled.

As demonstrated by the industry surveys discussed in Chapter 2, materials science and engineering is seen as critical to a wide range of technologies and industries. It is an enabling technology that permits or leads to advances in areas as diverse as aerospace, computers, communications, and automobile technology. In all these vital areas there is growing competition, and our major trading partners are catching up to or exceeding U.S. capabilities in the production of many materials and materials systems, that is, in the development of manufacturing technology. The principal driving forces in these competitive markets are specific industrial businesses rather than governments, but in general, coordinated government-sponsored R&D efforts can have a significant impact on industrial capabilities to compete. Notable examples in the history of U.S. development illustrate this impact; federal funding focused on aerospace-related R&D sponsored by the DOD and NASA and carried out in universities, government laboratories, and industry has been highly influential in the development of U.S. eminence in commercial aircraft manufacturing. Leadership in science does not guarantee leadership in engineering or technology. Cooperative mechanisms, fostered by government involvement, are being used increasingly throughout the world to enhance technology development and industrial competitiveness.

The complexity of modern manufacturing has led inevitably to interdependence among industries. This trend is on the upswing and is taking the form of joint ventures, licensing and use of outside sources for manufacturing via long-term contractual agreements, and increasingly, cooperation in the long-term R&D of technologies for improved manufacturing capability. In Japan, such cooperation, which is very advanced, is mediated by government laboratories in collaboration with industry. In the United States, the earliest examples of such industrial cooperation in precompetitive research can be seen in the funding efforts of such industry-sponsored research-granting organizations as the Electric Power Research Institute, the Gas Research Institute, and the Semiconductor Research Corporation; in R&D laboratories such as those of the Microelectronic and Computer Corporation; and in numerous industry and university centers. Noticeably lacking in the United States, and found to a greater degree in all the other countries studied for this report, is a national agency charged with stimulating and assisting industry and, when appropriate, with ensuring that cooperative activities are coordinated and that their impact on industrial development is optimized.

The recognition of materials science and engineering as a subject for focused national support is common to all the nations surveyed, but the organizational structure and funding mechanisms are as varied as the cultures and governments of those nations. There are, however, some important features to be noted in comparing the United States with Japan, and to a lesser degree, with West Germany. Japan and West Germany have been enormously successful in recent years in converting innovative concepts into technological advantage because of several factors operating in these countries: (1) education has been focused strongly on engineering rather than on science; (2) coordinated planning of targeted industrial development is stimulated by a government whose policies and expenditures are aimed at fostering the competitiveness of private industries; and (3) national laboratories are specifically charged with service to industry as a significant component in the complex process of transforming innovation to practices and products. These laboratories have almost no counterpart in the United States, because, with the exception of NIST, this country's national laboratories are not charged with the mission of service to the commercial sector.

APPENDIXES:
ISSUES IN MATERIALS RESEARCH

A

Synthesis

OVERVIEW

The term *synthesis*, as used in this appendix, refers to that part of synthesis and processing in which attention is focused primarily on the chemical and physical means by which atoms and molecules are combined to produce materials. In this view of synthesis, both the scientific problems of finding new routes to synthesis and the technological problems of finding ways to synthesize materials in useful quantities and forms are included. Thus the term *synthesis* as used here has a strong component of basic science.

The principal findings of the committee concerning the nature of this field and its current status in the United States are the following:

- Both intellectual opportunities and technological needs provide motivation for innovative research in the synthesis of materials.
- Synthesis underlies much of the progress in materials science and technology. Advances in synthesis yield new materials having new properties. Advances in synthetic techniques also may be used to enhance the properties of known materials by improving control of structure and composition. The physical phenomena that emerge in new or enhanced materials often lead to new technologies. Thus the health of U.S. technology requires active, aggressive, exploratory programs in materials synthesis.
- Close interactions between scientists and engineers engaged in materials synthesis, characterization, processing, and, ultimately, manufacturing are highly desirable in order to achieve rapid economic exploitation of research in synthesis.
- While U.S. excellence in molecular synthesis in areas such as medicine

and agricultural chemicals is clear, a comparable excellence in the synthesis of materials is lacking. In some areas involving materials synthesis, such as biotechnology, the United States still enjoys a leading position that needs to be preserved and strengthened. In other areas, such as polymer synthesis and solid-state chemistry, however, the United States clearly has fallen behind.

• Because of weaknesses in materials synthesis, there is growing concern that the United States is losing its dominance in basic materials research. In the past, U.S. scientists have been able to make up for weakness in synthesis by maintaining greater strength in characterization and analysis of physical phenomena. This system is no longer working, however, because foreign scientists with superior capabilities in synthesis are increasingly able to bring their own modern methods of characterization and analysis to bear on the study of new materials.

• Although many U.S. universities have strong academic programs in chemistry, materials science, and physics, very few have programs in solid-state synthesis, and fewer still have programs in which synthesis is strongly coupled to characterization and analysis. Moreover, academic materials research groups too often specialize in activities that require only narrow educational backgrounds.

• The decline of basic research in U.S. industry and the growing tendency to focus primarily on short-term product development in industrial laboratories have had an especially adverse effect on national capabilities in materials synthesis.

Additional findings of the committee are as follows:

• New educational experiences for many scientists and engineers will be required if the materials research community is to be able to carry out tightly coupled programs in synthesis, characterization, analysis, and processing. Both undergraduate and graduate academic curricula need to be reexamined with the goal of helping students achieve a level of basic understanding of materials synthesis and materials properties that will allow them to participate in fruitful exchanges of research ideas. Postdoctoral fellowships to encourage interdisciplinary research and, particularly, to encourage involvement in materials synthesis should be provided for talented students currently in doctoral programs in chemistry, physics, and materials science. All of the major providers of support for basic research should participate in funding such fellowships.

• Basic research in synthesis of materials should be increased in universities. Programs in polymer synthesis and inorganic solid-state chemistry, in particular, seem too few in number, especially in view of the wealth of U.S. talent in physical and organic chemistry.

• Scientists and engineers should be supported and encouraged to carry out interactive research in materials synthesis coupled with characterization and analysis of the properties of novel materials, and also coupled with the processing of materials. Because materials synthesis is intrinsically "small science," this is an area in which it should be possible to find funding modes that will encourage interactive research while preserving the flexibility and creativity that are associated with projects led by individual principal investigators.

• The U.S. system of federal and state-supported laboratories provides a framework within which it is possible to develop programs in synthesis that are closely related to technologically significant materials problems. This capability should be exploited more effectively than it has been in the past. In particular, federal laboratories might serve in part as resource facilities for the preparation of special materials to be used in R&D projects.

• Industrial corporations that depend on materials in order to compete successfully for world markets should recognize the benefit of maintaining in-house capabilities in materials synthesis. The long-term economic advantage of having ready access to superior materials usually justifies investment in research in the synthesis of new materials, the modification of existing materials, and the development of efficient synthetic methods.

Present U.S. corporate strategy, however, appears in most cases to favor withdrawal from long-range research, including research in the synthesis of materials. This general decline of industrial research has posed a dilemma for this committee throughout its various attempts to formulate meaningful recommendations, and the difficulty seems particularly severe with regard to research in synthesis. On the one hand, the committee strongly believes that research directly related to the synthesis of commercially useful materials can be performed successfully only in an industrial environment. On the other hand, most of the economic and political causes for the decline of industrial research are well beyond the scope of this report. Accordingly, the committee's position is simply that focused but flexible efforts, such as those suggested in the summary chapter of this report, should be made to help industrial organizations carry out research in materials synthesis, and that universities and national laboratories should participate in these efforts.

THE MEANING OF SYNTHESIS

As used in this appendix, the term *synthesis* refers to the chemical and physical means by which atoms and molecules are combined to form materials. Synthesis is the heart of the science of chemistry. It is the activity in which new materials are produced and in which new paths for the manufacture of materials are invented. It is the major source for the discovery of new chemical and physical phenomena in solids. Synthesis is a crucial component in the development of new technologies and in the improvement of existing technologies.

Synthesis, processing, and manufacture of solid materials form a continuum of intellectual and technological activities. The boundaries between these three elements of materials science and engineering are becoming increasingly blurred as the need for strong coupling between basic research and applied technology becomes more and more clear.

A practical reason for this blurring is the frequent need to choose synthetic reactions in such a way that products are suitable for further processing. For example, in the preparation of ceramics, it is generally desirable that the synthetic reactions produce precursor particles that are small enough to be compacted easily and sintered at relatively low temperatures. Another reason for the blurred distinction is that modern techniques for preparing materials often are direct combinations of what we conventionally have considered to be separate operations of synthesis and processing. A pertinent example here is the reactive injection molding of polymers in which the synthetic chemical reactions occur simultaneously with the "process" of molding. In this appendix, the focus is on activities that seem to the committee to fall primarily in the category of "synthesis." Discussion of topics that are primarily "processing" is saved for the appendix with that title. The reader should keep in mind that this distinction may sometimes appear to be artificial.

HISTORICAL BACKGROUND

Most of the major innovations in materials technology of the past two centuries have depended directly on the development of novel synthetic routes to existing or new materials. For example, in the early part of the nineteenth century, metallic aluminum was a curiosity available only in very small quantities. With the new synthetic route from cryolite developed by Hall, aluminum became a major commercial material. The field of synthetic polymers had its beginnings at the turn of the twentieth century with the development by Baekeland of phenol formaldehyde resins. However, the real revolution in this field began in the 1930s with synthesis of thermoplastics such as nylon, polyethylene, and polyester, which could be processed directly into film, fibers, or molded plastics. The plastics revolution, which has influenced every aspect of our lives, was made possible by novel synthetic approaches to both the monomers and the polymers. Thus synthesis not only has led to new and novel materials but also has provided low-cost alternatives that have made products affordable.

Rapid growth of the chemical industry in the 1950s and 1960s was so dependent on synthesis of new materials that most chemical companies maintained R&D centers consisting primarily of synthetic chemists. However, starting in the mid-1960s, some profound changes have occurred in U.S. industrial research that have altered sharply the way materials-related industries can respond to competitive challenges or generate new opportunities.

A number of industrial organizations—the metals industry is a prime example—have cut back on R&D to the point where little or no basic work is pursued. In these cases, the potential for improving antiquated commercial processes through new synthetic routes has become nonexistent.

In the area of ceramics, a number of important new directions have emerged in the past 20 years that suggest real opportunities, for example, stabilized zirconia, ceramics derived from sol-gel techniques, new types of refractory fibers, ceramic composites, and, most recently, the high-temperature superconductors. However, for the most part, the traditional ceramics industry does not possess the research capabilities necessary to pursue development of these novel concepts.

Some materials-dependent companies in the United States do recognize the long-range value of research in the synthesis of new materials and devote significant resources to such programs. Unfortunately, there are too few examples of this kind to be able to report a positive trend.

The history of programs in materials synthesis at U.S. universities, with one or two striking exceptions, parallels the weak effort in industry. Indeed, the lack of industrial job opportunities in this area has been a disincentive for academic programs. Conversely, the lack of academic programs and the resulting national shortage of materials scientists trained in synthesis have made it especially difficult for industrial laboratories to start new research projects in advanced materials. The level of activity in materials synthesis has remained low in academic institutions despite the fact that many of the leading research-oriented universities traditionally have maintained strong efforts in organic molecular synthesis. U.S. academic institutions even have been weak performers in the synthesis of polymeric materials. In the committee's opinion, there are fewer than 10 innovative polymer synthesis programs in U.S. universities.

Synthesis of inorganic solids is in little better shape. In the area of thin-film microelectronic materials formed by chemical vapor deposition, much of the chemistry appears to be done by electrical engineers who are interested primarily in the device characteristics of the materials being produced. While this situation is not necessarily bad, the lack of special expertise in synthetic chemistry leads to dependence on standard chemicals that can be acquired from commercial sources and standard methods of materials preparation. It seems likely that really new approaches to the design of materials systems will require novel synthetic techniques.

There are some signs of change in the academic scene, but there is also much inertia in the system that will hinder rapid strengthening of programs in materials synthesis. Part of the difficulty is that people who are creative in synthesis often are not aware of the needs and opportunities for new materials. Further, the culture of chemistry has been such that a synthetic chemist could make new materials, albeit molecular materials, and fully characterize them without ever going beyond the boundaries of a conventional chemistry department. The characterization of new solid materials, on the other hand, generally requires a much broader range of instrumental capabilities and scientific expertise. In short, modern materials research is necessarily an interactive enterprise involving many scientific specialties. The fact that synthesis is an absolutely essential part of this picture is the principal message of the next paragraphs.

THE ROLES OF SYNTHESIS IN MATERIALS RESEARCH

It is useful to distinguish three separate roles that are played by synthesis in materials research:

1. The synthesis of new materials.
2. The synthesis of known materials by new techniques or in novel forms.
3. The synthesis of known materials for special purposes, for example, for use in research or in technological development.

The historic weaknesses of U.S. institutions in materials synthesis have caused problems to emerge in connection with each of these roles. These problems will be discussed in the order listed above.

1. Research in the synthesis and characterization of new materials has been demonstrated repeatedly to be the basis for the discovery of new phenomena and for the development of new technologies. Since about 1970, the United States has relied largely on foreign laboratories for the synthesis of new chemical compounds. The common experience was that foreign scientists would publish compositions and structures but would not be very quick to study the properties of novel materials. Subsequent discoveries of new phenomena usually occurred in the United States. Examples of discoveries or technological developments that have taken place in this way include charge-density waves, conducting polymers, high-field superconductors, intercalation compounds, and high-dielectric-constant microwave resonators.

This informal arrangement—"they" discover the materials and address questions of the existence and stability of particular structures, and "we" then discover new phenomena and properties—has never really been satisfactory. In the first place, it requires that U.S. scientists be able to prepare the new compounds themselves by following the published recipes. At past levels of funding, even that step was only marginally possible, with much of the best work being done only in industrial laboratories such as AT&T Bell Labs, Du Pont, or IBM. Indeed, many scientists involved in materials research could not obtain samples of desired substances for study and were not able to prepare them themselves.

More importantly, foreign scientists are increasingly able to search for new phenomena and new properties of materials in their own laboratories—often the same laboratories in which the materials being studied originally were synthesized. In this way, they are often able to capture both the scientific excitement and the technological advantages of their research.

The recent discovery of high-temperature superconductors is a case in point. These materials were first synthesized in France, and their superconducting properties were discovered in Switzerland (at an IBM laboratory). Much of the best work in this field is now being carried out in Japan, where scientific leaders were early to recognize the need for strong coupling between synthesis and other areas of materials research, and have acted to meet this need by supporting appropriate programs at universities and at industrial and state-supported laboratories. In each country's laboratories, synthesis capability and the study of materials properties are linked. The lack of similarly strong programs in the United States will lead to further decline in our scientific leadership and technological capabilities. Consequently, for the future health of all of materials science and engineering, a much stronger focus on the synthesis of new materials seems urgent.

2. New techniques are needed for the synthesis of known materials that are useful either for scientific or for technological purposes but that cannot currently be prepared in the proper form, in adequate quantities, or at reasonable expense. Examples of needed techniques in this category include efficient methods for preparing advanced ceramics and special polymeric materials. In addition to the development of methods to address already known needs, basic investigations of novel techniques need to be supported in order to provide a foundation for future developments. Such investigations might range from fundamental studies of metastable solids to exploration of the effects of various methods of preparation on the stability and lifetime in service of materials used in electronic devices.

The motivation for developing new methods for synthesizing known materials is primarily technological, and therefore such research is best performed by industrial organizations that understand their own goals and constraints. As mentioned above, however, U.S. industry has largely abandoned research of this kind. Current trends in federal support for materials science and engineering—note, for example, the National Science Foundation's rationale for science and technology centers—imply an expectation that some of the necessary effort in technologically important areas such as synthesis might be taken up by research groups at universities.

3. The synthesis of known materials for special scientific or technological purposes is essentially a service function. For example, solid-state physicists and electronic engineers depend strongly on the availability of carefully prepared samples of the materials that they are using. Sample preparation, however, is not a particularly challenging or rewarding activity for research chemists. Nevertheless, such synthesis often requires considerable effort and

expertise as well as expensive and well-maintained equipment. Research-grade samples generally cannot be purchased commercially because the cost of their preparation seldom can be recovered on a direct-charge basis. Yet the phenomena to be discovered might be scientifically and technologically very exciting.

Because there is an obvious need for special materials to be used in R&D, it seems very much in the national interest that some mode be found for the support of laboratories that will perform this kind of synthesis. Perhaps some part of this function could be served by government-related laboratories if funding were available for that purpose. But a satisfactory solution to the problem of availability of research-grade materials can come only as a part of a broader effort to upgrade the status of materials synthesis in all of the nation's research institutions.

NEEDS FOR MATERIALS SYNTHESIS: SOME EXAMPLES

In the remaining parts of this appendix, the various roles played by materials synthesis are illustrated by selected examples of needs and opportunities for research in this area. In this section, some needs that arise in commercial and defense-related applications are discussed. In the next section, some opportunities of a more fundamental nature are described.

Information Industry

The large-scale need for synthesis of new polymeric materials is a relatively recent development in the information industry. Until a few years ago, the only places where polymers played key roles in the manufacture of large mainframe computers were in lithographic processes for patterning metal lines and in the bindings of the particulate disks used for magnetic storage. In the 1980s, this picture has changed to the point where one can predict that future progress will depend on the availability of polymers that do not exist today. These needs extend throughout the entire hierarchy of computer hardware, including semiconductor chips and packages, optical and magnetic recording systems, and printers and displays such as electrophotography, impact printing, ink jet printing, and liquid crystal displays. For illustrative purposes, attention is focused here on the needs for new polymeric synthesis in the packaging of electronic components.

An important driving force in packaging technology is the rapid growth in the capacities of RAM chips from 50 kilobits just a few years ago to 1 megabit today and to a projected 64 megabits by the mid-1990s. With higher device densities, problems of wireability and reliability become more critical. A whole new set of materials needs emerges for these advanced structures.

The present trend is to build more of the circuitry directly into the chips and to package many interconnected chips within single modules, cards, or boards.

One can readily translate this trend into specific needs for new materials. Increasing the density of devices on chips requires development of submicron metal lines and an increase in the depth of circuitry from two or three levels to five or six. These developments will require, respectively, new photoresists with strong sensitivities at short wavelengths, and insulating polymers that planarize and have unusually low dielectric constants. Increasing the complexity of the package causes problems of reliability associated with mechanical stresses at the interconnections between the various devices. These stresses arise from the mismatch in thermal expansion coefficients between the silicon of the chip, the alumina of the module, and the epoxy fiberglass in the card or board. Reducing these stresses requires development of polymeric substrates for modules, cards, and boards that better match the thermal expansion coefficient of the silicon chip.

Transportation Industry

Several needs for synthesis of advanced materials are common to both the aircraft and the automobile industries. For example, lightweight structural materials are needed for fuel efficiency. Significant progress has been made over the past 15 years in the development of lightweight graphite-fiber-reinforced epoxy composites. The synthesis of tougher matrix resins might allow such composites to be used more widely in structural elements such as aircraft wings. Graphite fiber composites are relatively expensive for use in automobiles, and thus the primary emphasis has been on fiberglass-reinforced resins. In the future, we might look for the synthesis of inexpensive, self-reinforcing, injection-moldable liquid crystalline polymers to bring about major advances in structural materials for automobiles. In fact, certain types of liquid crystalline polymers retain mechanical properties up to about 350°C and might even be used in parts of engines.

Another need in the transportation industry is for improved flame-resistant materials for use in the interiors of aircraft and automobiles. Currently, too little effort is directed at designing new kinds of textiles that are intrinsically flame resistant and that do not produce smoke or toxic gases in a fire. The challenge is not only to achieve the key feature of flame resistance, but also to provide comfort and wearability, and to make such textiles available at costs competitive with those of currently used materials.

The search continues for inexpensive, high-temperature ceramics that can be used to increase the efficiency of combustion engines. The intrinsic brittleness of ceramics remains a major stumbling block, notwithstanding recent progress in the preparation of composites reinforced with stabilized zirconia and fiber-reinforced composites. Success in this area will require synthesis of ceramic materials that can be processed into finished shapes quickly and cheaply.

Many needs for new materials arise in the design of advanced supersonic aircraft. The development of lightweight structural skins that retain their strength at 250°C will be necessary for aircraft designed to fly at Mach 3. Even more formidable are the materials requirements for the national aerospace plane, which is expected to have surface temperatures in the engines and leading edges that are in excess of 1500°C.

Energy Industry

A number of our present energy technologies could be improved in efficiency or safety through the synthesis of new materials. The success of technologies like solar energy conversion or automobiles powered by fuel cells or batteries will depend on whether new materials and economically viable ways of making them can be discovered. Unfortunately, much of the momentum of research in these areas has been lost in recent years.

One example of the role of new materials in what may be a future energy technology is associated with fuel cells. Fuel cells are devices for converting chemical energy directly to electricity with high theoretical efficiency, with no moving parts, and without the pollution that accompanies combustion. The advantages of fuel cells seem evident, but the development of a technology based on them is limited by the materials available for use as electrodes. For example, there is no fuel cell capable of directly consuming liquid fuel because there is no known electrode at which such fuels can be oxidized at useful rates. Development of new electrode catalysts for fuel cells might lead to dramatic improvements in the efficient use of energy resources and could help solve pollution problems.

The energy industry also has a need for synthesis of passive materials. For example, the nuclear-based technologies need new, highly nonreactive materials for waste storage, and they also need to improve techniques for suppressing corrosion in cooling systems. Achieving superior materials properties in these cases will require the sort of materials synthesis in which the United States has not been strong. Since the nuclear industry is currently in a state of flux, it is unlikely that materials synthesis will receive a great deal of attention.

Solar energy conversion has been proven theoretically to be feasible in the sense that direct conversion of light to electricity with efficiency greater than 15 percent has been achieved in the laboratory. It is evident, however, that large-scale generation of electricity directly from sunlight will not be possible until inexpensive and efficient photovoltaic technology becomes available. At present, thin-film materials look the most promising. There is already a profitable photovoltaic industry in Japan, but, even there, large-scale energy production has yet to be realized. Solar energy technology seems a good bet for the future, but advances in the synthesis of photovoltaic materials will have to precede further development.

Like solar energy, nuclear fusion looks promising in the long run. In this future technology for central power generation, the materials problems are severe and somewhat less well defined than in the photovoltaic area. Some of the problems of conventional nuclear energy will be encountered here, but the most challenging research is likely to be focused on the development of structural materials that will survive the extremely harsh conditions expected in the fusion reactor. For the most part, such materials do not exist today, and future success in this area will be determined by the ability of scientists to come up with innovative solutions.

Defense and Aerospace Industries

The advanced materials needed for defense and for aerospace have very special purposes. They often do not have the commercial applications that would justify support for their development in the private sector. Thus government agencies must take the lead in supporting synthesis of novel materials for military applications. Outstanding needs include systems having unique optical and electronic properties as well as systems having unusual strength and toughness.

The materials needed for defense and aerospace span such a wide range that only a few possibilities for materials synthesis in this area can be described here. One such possibility is the development of multitechnology chips that combine infrared detectors with electronic and optical processors. The technical issues here include problems of synthesizing each of the component materials in the presence of the others without destroying the function of any of them.

The Department of Defense has unique needs for new materials in developing its "low observables" or "stealth" technology. The synthesis of such materials requires improved understanding of the physics of detection and of the properties of materials that will be needed to defy detection. The interplay between synthesis, processing, and physics in this area must be strong, especially if the resulting materials are to maintain structural properties with minimal increase in mass and volume.

The Department of Defense also needs unusual materials for sensing and surveillance. Applications range from tracking submarines to detecting toxic chemicals. For example, it would be very useful to find a material superior to polyvinylidenefluoride as an element in acoustic detectors. Electronic ceramics, redox polymers, biomaterials, and membranes are all likely to play roles in advanced chemical sensing systems.

These few examples illustrate the diversity of defense- and aerospace-related materials problems that might be solved, at least in part, by synthesis. The needs for defense are as wide ranging as those in the private sector. The difference is that the military needs will be the sole responsibility of government because, for the most part, commercial markets for advanced military materials are limited. However, the innovative research in synthesis that will be necessary for meeting these needs is not currently under way in academic, government, or industrial laboratories.

OPPORTUNITIES IN SYNTHESIS

Synthesis provides scientists and engineers with the raw materials for their work, whether that work be fundamental or applied. The wide variety of technological needs described briefly in the preceding paragraphs tend to "pull" research in synthesis. At the same time, materials synthesis is "pushed" by advances in science and engineering. This appendix concludes with a few examples that seem to fall in the latter category of opportunities for research.

Ultrapure Materials

New synthetic methods and new procedures for handling materials during synthesis are now yielding substances of unprecedented purity and performance. The synthesis of very pure substances is becoming increasingly important both in microelectronics and in the development of new structural materials.

The need for extremely pure silicon in microelectronic devices is well known. Currently, very pure and atomically perfect III-V semiconductor crystals are being used in advanced electrooptical devices. The production of volatile precursors (SiH_4 for silicon, AsH_3 and gallium alkyls for GaAs) in 99.9999 percent purity presents a challenge to the engineers who are developing the technology for producing these ultrapure semiconductors. Similarly, the production of actinide-free ceramics for packaging radiation-sensitive very large scale integrated circuitry requires reengineering the synthetic technology.

In the area of structural materials, it is becoming clear that oxygen and carbon impurities can limit the strength of fibers used in composites. Crack initiation and growth in solids can be attributed to impurities and defects. Thus, as in the case of microelectronic materials, the ability to synthesize extremely pure ceramics—and to do this in economically significant quantities—will determine the success of this technology.

The success of molecular precursors in solid-state synthesis often depends upon the use of ultrapure molecular materials. For example, a promising new method for producing ceramic fibers starts with the synthesis of a pre-ceramic polymeric material that can be processed into a fiber and then pyrolyzed to form the ceramic. The purity of the preceramic polymer turns out to determine the strength of the ceramic fiber. In another example, very pure organometallic precursors can be used in synthesizing complicated ternary and quaternary III-V compounds for use in the preparation of device-quality materials. Lasers with unusually low threshold currents have been produced in this way. Finally, molecular precursors to metal lines and thin films may be useful in device fabrication. In all of these examples, synthetic chemistry must be integrated with the design of devices and the engineering of production processes.

Organic nonlinear optical materials provide another illustration of the opportunities for research in the synthesis of ultrapure substances. It is becoming widely appreciated that the nonlinear optical properties of organic and organometallic molecules can be superior to those of inorganic solids. Preparation of new organic and organometallic molecular solids for use in nonlinear optics represents a special opportunity for academic chemists because the synthetic techniques are relatively commonplace in major research-oriented chemistry departments in the United States. What makes this a new opportunity is the need to prepare organic and organometallic solids of a purity and optical quality seldom demanded in typical chemical applications. New strategies for designing and preparing organic and organometallic solids with good optical properties are needed. As in the preceding examples, practical applications of new optical materials will require technologies for producing the solids in useful forms—monolithic crystals, crystal core fibers, oriented films, and the like. Synthesis will have to be integrated with fabrication, as discussed in the paragraphs on shape-limited synthesis below.

One of the most critical and pervasive needs for ultrapure materials is in fundamental physical studies of structure-property relationships. For example, very pure samples of polymers with narrow distributions of molecular weights are needed in order to test modern theories of the behavior of polymers in dense fluids. Thus, for both fundamental and applied purposes, laboratories dedicated to the preparation of ultrapure materials must have high priority in programs aimed at upgrading U.S. capabilities in materials science and engineering.

Shape-Limited Synthesis

The term *shape-limited synthesis* refers to a relatively new approach to the preparation of materials—a combination of both synthesis and processing—in which one of the chemical reactants is formed from the beginning in the shape of the final product. In more conventional technologies, new materials first are synthesized in whatever form emerges from the reactor—often a powder—and then processed by various methods—e.g., molding and pressing—to achieve the desired shape. This approach requires that the synthesized material not only possess a set of desired properties, but also be suitable for later-stage processing. This latter requirement has constrained scientists to focus much of their attention on those materials that can be processed using commercially available equipment. This constraint has not encouraged innovative materials synthesis.

The idea of chemically converting a precursor body into a new composition while retaining the original shape first was demonstrated by the preparation of fibers with compositions that had been inaccessible by traditional fiber-forming methods. Thus fibers of boron nitride were prepared by nitriding fibers of boric oxide, and cross-linked phenolic fibers were formed by reacting Novolac resin filaments with formaldehyde. One variation of this approach is to incorporate into the precursor fiber all of the necessary reactants. For example, silicon carbide fibers can be prepared directly by the pyrolysis of a carbosilane fiber. For many ceramic processing steps, it is desirable to have monodisperse, submicron, spherical particles. One of the most practical applications of sol-gel technology is the preparation of spherical ceramic particles with precisely controlled sizes. Similarly, the engineering of new gas-phase processes for synthesizing polyolefins can make possible the production of dry, flowable microspheres of polypropylene and high-density polyethylene.

Chemically modifying surfaces to achieve surface passivation of sensitive materials is another version of shape-limited synthesis. For example, in the processing of silicon into semiconductor chips, the surface of silicon can be very thinly oxidized to protect the active devices during subsequent processing. A similar procedure would be highly desirable for passivating gallium arsenide surfaces. For the future, the ability to control surface chemistry will be essential in the design of devices with nanometer dimensions.

The versatility of the shape-limited approach for preparing new fibers or thin coatings is considerable. On the other hand, to prepare larger shapes, it is essential to minimize changes in density between the starting material and the product. Such changes may produce stresses that cannot easily be relieved in this kind of process. Innovative techniques for overcoming this limitation would be most desirable.

New Synthetic Methods

At present, too little research is being carried out in the United States on truly novel synthetic methods. Most current research on synthesis emphasizes the use of conventional techniques for tailoring structures of molecules to achieve specific properties. Few completely new synthetic approaches are being developed, and, as a result, U.S. scientists are becoming comparatively limited in the techniques they can bring to bear on problems of synthesis.

Perhaps even more important, opportunities for unexpected discoveries are being lost.

The committee believes that there is no lack of good ideas to be explored. For example, laser-assisted chemical processing is a new approach for synthesis of metals as films or deposited wires. It has recently been demonstrated that metals such as copper, gold, chromium, and cadmium can be thermally deposited with laser assist onto surfaces by using the appropriate organometallic precursor. Similarly, relatively cool plasmas are being used to pyrolyze hydrocarbons into carbon radicals that deposit as diamond films. The engineering of industrial-scale plasma processes presents major challenges. With hot plasmas, it may be necessary to cool the gases at rates of millions of degrees per second in order to trap the thermodynamically unstable species that are necessary for synthesis.

There is also an opportunity for stronger efforts aimed at improving catalytic processes. Major progress is being made these days in understanding the structures of surfaces and how they function as catalysts. The ability to modify such surfaces chemically at the angstrom level—for example, via molecular beam epitaxy or atomic layer epitaxy—offers opportunities for the design of completely new types of catalytic systems.

A final example of novel synthesis concerns solid-state preparation techniques. The most commonly used methods involve reacting mixtures of powdered substances at high temperatures. Low-temperature methods, such as the increasingly popular sol-gel techniques, not only are more efficient but also allow the preparation of many new phases that are unstable at high temperatures. Again, there is no lack of opportunities for innovation. But innovation cannot occur unless there is financial support and institutional encouragement for research in synthesis.

B

Processing

OVERVIEW

The processing of materials is crucial for achieving quality and efficiency in any technological enterprise. Processing, like synthesis, deals with the control of atoms and molecules to produce useful materials. However, processing also includes control of structure at higher levels of aggregation and may sometimes have an engineering aspect. The way in which materials are transformed—that is, "processed"—to form bulk materials, components, devices, structures, and systems is a major factor in determining the success of efforts in areas as diverse as the electronics and construction industries. Competence in materials processing is essential for the conversion of new materials into successful products, and for the continued improvement of products made from currently available materials.

Today's needs and opportunities for research in the processing of materials call for contributions from, and close interactions between, universities, industry, and the national laboratories:

• An integrated and interdisciplinary approach to education in materials processing is required in order to take advantage of new opportunities. The role of materials processing as an integral part of manufacturing technology has not received adequate emphasis in academic curricula, probably because it draws upon several disciplines and is a complicated mix of science, engineering, and empiricism.

• The development of commercial materials processing technology is the special and indispensable province of industry. A strong industrial capability for carrying out research relevant to materials processing will be essential for achieving high quality and cost effectiveness in all industrial sectors.

- The national and federal laboratories, with their extensive technical resources and special facilities, are well positioned to make major contributions to research in materials processing.

Recent scientific and technological developments have provided a remarkably diverse range of challenges and opportunities for research in the processing of materials. Among these opportunities are the following:

- *Process modeling and simulation*: High-speed computing capabilities coupled with improved theoretical understanding provide opportunities for improvements in materials processing technologies.
- *Ceramics*: Improvements in the technologies for processing of ceramic materials will be required in order to fabricate the new high-temperature superconductors, as well as to exploit the potential of ceramics in a broad range of electronic and structural applications.
- *Optoelectronic materials*: Advances in an array of processing technologies will be necessary in order to realize the opportunities present in optoelectronics. These technologies include crystal growth, molecular beam epitaxy, and physical and chemical vapor deposition.
- *Rapid solidification*: The continued development of this processing technology should lead to a wide range of applications and products.
- *Metals*: The processing technologies of the metals industries require a major infusion of resources, with an emphasis on areas such as automation and recycling of materials.
- *Polymers*: Processing techniques such as reaction injection molding, melt spinning, and polymer forging offer significant potential for further development.

Technological challenges in the processing of materials, such as those listed above, bring with them needs for improved understanding at more basic scientific levels. Areas in which new needs for fundamental research are emerging include the following:

- *Interfaces*: Much of materials processing consists of manipulating and controlling the interfaces that separate various components of complex substances. New experimental methods and characterization techniques should lead to significant advances in understanding interfaces.
- *Nonequilibrium materials*: Materials processing technologies generally deal with materials in states that are very far from thermodynamic equilibrium. Experimental and theoretical advances in understanding the thermodynamics and kinetics of nonequilibrium states are now within reach and should lead to improvements in processing technology.

MATERIALS PROCESSING AND ECONOMIC COMPETITIVENESS

The compelling theme of national economic competitiveness leads directly to a strong focus on materials processing. Processing and fabrication of materials—that is, the transformation of raw materials or synthesized substances into components, products, devices, structures, or systems—is an essential activity throughout U.S. technology. This is true for industrial sectors as diverse in nature and size as construction, computers, steelmaking, aerospace, electronics, and transportation.

The term *materials processing* refers to an enormously wide range of technologies. A very incomplete list includes refining of metals, rolling of sheet steel, shaping of metals by machining, thermomechanical processing of alloys, growth of gallium arsenide crystals, zone refining of silicon, pressing and sintering of ceramic powders, ion implantation in silicon, formation of artificially structured materials by molecular beam epitaxy, spinning of high-strength polymeric fibers, gradient doping of glass fibers, sol-gel production of fine ceramic powders, modification of concrete by addition of polymers, and lay-up of composite materials. Some of these processing technologies are quite new and may, in time, lead to major technological advances and industrial growth. Others are well embedded in established industries but require continual improvement to maintain competitive positions.

The nation's eroding competitive position in some industries stems largely from national weaknesses in materials processing technologies. The essential difficulty seems to be that, with increasing frequency, industries abroad are bringing the results of R&D, or of innovative design, to production and thence to the marketplace in considerably less time than is the present practice in this country. In too many instances the new products produced elsewhere are more innovative in design and more reliable in function, and are produced at a lower cost than in the United States. The Japanese in particular have developed a formidable capability in manufacturing and the related materials processing technologies in a diverse set of industries, ranging from steelmaking, automobiles, and shipbuilding to dynamic random access chips and video cassette recorders. Other nations, such as South Korea and Brazil, are now following the same path.

For about four decades, the United States has been accustomed to holding a dominant position in virtually every important industrial sector. This situation was one of many inheritances from the Second World War, and it could not have been expected to continue indefinitely. Technological developments in transportation and communications have accelerated the trend to a globalization of technology and trade. In the young and rapidly evolving international economy the proper balance between the technological strengths of the U.S. and foreign economies is not yet evident.

Although the nature, diversity, and scale of materials processing technologies differ markedly among industrial sectors, there is a universal need to improve technological competence and efficiency in all aspects of materials synthesis, processing, and fabrication. Increased competence and sophistication in manufacturing and the accompanying materials processing technologies are necessary to produce high-quality and reliable products in an efficient and cost-effective manner. Competence in materials processing is also essential to the efficient incorporation into technology of advanced materials such as structural ceramics, composite materials, and high-temperature superconductors.

HISTORICAL BACKGROUND

Materials processing is probably the oldest of human technological activities, extending back in time to far before the advent of recorded civilization. For many centuries, materials processing was an empirical activity. Progress depended upon careful observation, extensive trial and error, and, no doubt, shrewd insight. There was no guidance, however, from what would now be regarded as soundly based scientific understanding. Nevertheless, by this empirical approach, a substantial variety of materials—metals, alloys, ceramics, glasses, concrete, and so on—were developed in useful compositions and shapes for both functional and artistic purposes. It was not until the seventeenth and eighteenth centuries, when a sound understanding of both chemistry and mechanics began to emerge, that materials processing underwent a transition from a craft to a science-based technology. Iron processing and steelmaking were among the first areas to benefit from the new knowledge.

In present times, the infrastructure of our society has come to depend strongly on the existence of highly diversified and sophisticated materials technologies. The technological foundations of sectors such as agriculture, transportation, manufacturing, communications, construction, and national security are dependent upon the ability to process materials into a variety of compositions and shapes and to do this with accurate control of properties. An enormous proliferation of diverse materials—from stone and bronze, to steel and ceramics, and, most recently, to composites and high-temperature superconductors—has been driven by the demands of our technological infrastructure. The associated technologies for processing these materials are equally diverse. The growth in the materials base has been explosive in the decades following the Second World War, and there is every sign that this growth will continue well into the twenty-first century.

THE PRESENT SITUATION: SOME EXAMPLES

The processing of materials is an enabling activity, one that is essential to the practical realization of the ideas generated by scientists, engineers, and designers in all fields, and one that can greatly expand the potential for performance of materials.

In many instances, materials processing is a critical step in the development of an entire technology. Electronics is an outstanding example of an industry that depended from its beginning on the invention of processing techniques—in this case, the growth and purification of semiconductor crystals. In other instances, new materials processing techniques may lead to materials substitutions in existing technologies with striking improvements in performance and reductions in cost. Examples of the latter situation are the substitution of optical glass fiber for copper wire in long-distance communication systems and the replacement of some metallic components in aircraft and automobiles by composite materials. In all cases, materials processing draws upon the results of research and design and also relies upon empirical investigations and extensive testing programs.

Semiconductors

The birth and rapid growth of the electronics and computer industries required an extensive series of interacting developments in basic research, materials processing, and design. An early and key step was the invention of the transistor in the late 1940s. This has been celebrated, and justly so, as a triumph of basic research. The enabling fundamental studies were strongly knowledge driven, although they were carried out in the context of a search for radically new communications technologies. In the course of the basic studies, both experimental and theoretical, wholly new patterns of electronic behavior were discovered and interpreted.

The invention of the transistor was the first step, albeit the most important, in a long series of discoveries and technical developments that were necessary to transform the discovery of electronic amplification in solids into the modern electronics and computer industries. Many of those still dynamic and evolving technologies are materials processing techniques that were developed specifically to exploit the new scientific discoveries for practical purposes. These were application- or market-driven developments, in contrast to the invention of the transistor, which was knowledge driven.

An excellent example of application-driven research is the development of the zone refining method for purifying silicon, which followed shortly after the invention of the transistor. This research was application driven in the sense that silicon and germanium semiconductors required impurity levels that were far below those obtainable by existing technologies at that time. In zone refining, purification is achieved by the passage of a thin molten zone, which removes the impurities, through the otherwise solid material. This invention was essential to the successful production of homogeneous high-purity crystals at reasonable cost.

A third major innovation was the design concept that culminated in the invention of the chip, that is, the circuit on silicon. The chip became the focus of integrated circuit technology, which required many new processes, such as patterning, photolithography, and metallization.

It is evident from this brief description that the birth of the electronics industry required major contributions from fundamental knowledge, applied research, and innovative design. The further explosive development of the electronics and computer industries has been paced by continuous improvements in the interactive technologies for silicon processing, circuit design, and manufacture of integrated circuits. For example, it has continued to be necessary to improve crystal growth procedures in order to provide larger substrates for production of chips. New techniques have also been developed for doping silicon with the appropriate electrically active solutes, first by diffusion processes and then, more recently, by ion implantation. These are but a few of the many developments in materials processing technology that have been required for the rapid growth of the electronics industry.

An important but often overlooked aspect of advances in materials processing is the development of new instrumentation for both processing and characterization. Examples from the electronics and computer industries include crystal growers, optical steppers, deposition systems, photolithographic and electron beam machines, and Rutherford backscattering systems for elemental analysis. Effective, state-of-the-art instrumentation is essential for improvements in the materials processing component of any industry.

Optical Glass Fibers

Just 20 years ago it was realized that glass fiber, as a replacement for copper wire, offers a considerable advantage as a communication medium at optical frequencies. Both improved performance and reduced costs were predicted. Following a period of intensive development of methods for processing glass fibers, the conversion of most long-distance communication in the United States to optical frequencies is now complete. An optical fiber cable has been laid across the Atlantic Ocean; conversion of the long-distance network in Europe from copper to glass is under way; and it is expected that distribution networks and short-distance links in the United States will be converted to optical fiber in the next decade. It is a matter of perspective whether this is regarded as merely the result of a materials substitution or as the development of a wholly new technology for the communications industry. It is certainly a major development in terms of improved performance and reduced costs.

The production of hundreds of thousands of miles of glass fibers of uniform properties suitable for optical transmission of information required development of a new materials processing technology. This technology included a sound choice of material composition, vapor deposition techniques for the elimination of optically absorbing impurities, careful chemical and mechanical design to eliminate optical leakage, and appropriate fiber-drawing techniques.

The important phenomena for optical transmission are scattering loss, absorption loss, and dispersion. In the new silica glass fibers, absorption loss has been eliminated by reducing the concentration of optically absorbing impurities to below 1 ppb. Propagation losses for fused silica fibers are now at an intrinsic minimum of approximately 0.1 dB/km, caused by Rayleigh scattering from density fluctuations. To achieve even lower losses, which would allow greater distances between repeater stations, wholly different glass compositions must be considered.

The development of optical glass fiber was application driven. It was not derived from new basic research, as was the case for the transistor, but rather from a careful engineering, systems, and economic analysis. The development of the new glass fiber processing technologies was carried out in industrial laboratories, with relatively little input from universities or the national laboratories.

Rapid Solidification Technology

Rapid solidification is an emerging technology. Its roots are in the universities and national laboratories as well as in industry, but most processing development currently is being carried out in industrial laboratories. There has been good interaction between the communities. It is probable that in the case of rapid solidification basic scientific progress has been accelerated because of the excitement generated by the potential for commercialization.

Rapid solidification is an advanced example of nonequilibrium processing of materials from the liquid state and therefore is the latest in a long line of metallurgical developments that probably started with the accidental melting and casting of native copper many centuries ago. Although there were several experimental and theoretical precursors to the technology of rapid solidification, the critical experiment was probably the demonstration in a university laboratory some 26 years ago that metallic glasses—noncrystalline solid solutions of metallic components—could be produced by extremely rapid quenching from the melt. This powerful demonstration of the utility of rapid cooling in producing unexpected nonequilibrium structures was shortly followed up in many university and industrial laboratories. It opened a new realm in the study of materials, microstructures, and properties.

In conventional or ingot processing, the cooling rates from the liquid state are of the order of 1 K/s or less, and may be as small as 10^{-3} K/s. The cooling rates in rapid solidification processing, however, may vary from

10^2 K/s to as high as 10^8 K/s. These rates are so high that thermally driven diffusive rearrangement of atoms during cooling is impeded or wholly prevented. Highly nonequilibrium structures are then produced, often with unique properties. At sufficiently high rates, many alloys can be produced with an amorphous, or glasslike, atomic structure. At somewhat slower but still quite high rates, very fine grained crystalline structures are obtained, often in composition regimes that are not accessible to conventional processing.

Very high cooling rates require rapid extraction of heat. Therefore at least one dimension of the solidified material must be small so that the whole sample can be in close thermal contact with a cold substrate. Consequently, rapid solidification technology cannot be used at present to produce large, monolithic objects.

Rapid solidification technology has led to materials with new and useful combinations of magnetic properties. Attempts now under way to exploit their unique soft magnetic properties should lead to applications in electronics, power distribution, motors, and sensors. New permanent magnets recently produced by rapid solidification should be useful in building compact, powerful, electric motors.

Direct ribbon casting is a version of rapid solidification that has much promise for producing thin sheets of materials with unusual combinations of properties. Among the present applications where this technology has led to improved performance and lower costs is the production of brazing filler alloys, solder alloys for electronic packaging, and thin stainless steel sheet.

Rapid solidification also has been used to produce fine-grained and homogeneous crystalline—as opposed to amorphous—materials with much-improved properties and performance. The materials that have responded well to this processing technology include high-strength aluminum and magnesium alloys, tool steels of high toughness, and nickel-based superalloys.

Rapid solidification recently played a key role in the remarkable discovery of the so-called quasi-crystalline phases. These phases were first produced accidentally during rapid solidification of aluminum-manganese alloys. The scientific interest in these phases arises from the fact that they display long-range order—they are not amorphous or glassy—but the symmetry of the order is not consistent with the heretofore accepted rules defining the allowable symmetries of crystals. The discovery of quasi-crystals has led to an ongoing reexamination of the basic principles of crystallography, a science that now will have to be reformulated in a more general framework. It is not known at present whether these new phases will have interesting and useful properties, but this entirely new phenomenon clearly calls for intense investigation. It is surely interesting and instructive that a study of structure and properties through rapid solidification processing should lead to a major discovery in crystallography.

Steelmaking

The production of iron and steel is materials processing on a wholly different scale from that of the production of the high-technology materials discussed in the previous examples. The annual production of most high-technology materials is in the range of thousands of pounds per year, whereas the current annual output of U.S. steel mills is about 70 million tons. Furthermore, the production rate in the steel industry is extraordinary in comparison with that in other materials industries. Such volumes and rates require process technologies that are, in important ways, as complicated and sophisticated as those that are found in the processing of high-technology materials.

For example, with a modern basic oxygen furnace, approximately 300 tons of steel are produced in about 30 minutes from molten iron containing 4 percent (by weight) carbon. Twenty minutes later, this steel is decarburized with vacuum degassing to less than 0.003 percent by weight carbon. In another 40 minutes, the steel is continuously cast into 9-in.-thick slabs. Such a complex materials processing technology requires a fundamental and detailed understanding of a whole series of complicated materials phenomena, including interactions in gas-metal, metal-slag, and metal-refractory systems. An equally deep understanding of fluid flow and heat transfer in complicated geometries over a wide range of temperatures is required. Process control is maintained by an array of computers operating on programs developed from detailed simulation and modeling of each step in the process. An expert system decides if the slabs produced meet the designated specifications. Further processing of the steel requires equally detailed modeling and process control technology. Still, continual and extensive process development will be required in the future to produce steel at a competitive quality and price.

Steel remains the most versatile of the materials of the technological infrastructure. In contrast to most ceramics and composite materials at their present stage of development, steels are tough and their properties are both predictable and reproducible to a high degree of precision. Steels can be processed to exhibit a wider range of useful combinations of mechanical properties than any other material. Steels are unique in their sensitivity and range of response to variations in process technology, including composition, mechanical deformation, heat treatment, and thermomechanical processing.

In recent years, however, the in-house capability of the U.S. steel industry for developing new technology has steadily eroded. This erosion is closely tied to the problems of competition and profitability that the industry has experienced. Some U.S. steel companies have improved their positions through the purchase of Japanese materials processing technologies, but that tactic is at best a catch-up game, one that cannot lead by itself to a position of technological strength. It is true that the purchase of technology from abroad was one component of the Japanese drive to technological leadership. However, the Japanese steel industry also put a high priority on developing an in-house capability for R&D and, as a consequence, was able immediately to absorb the new technology purchased from abroad and to effect further improvements in a timely and efficient manner.

An important but not often noticed consequence of the decline of the U.S. steel industry is that, in the area of materials processing, there has been a marked decrease in interactions between the industry and the universities. This situation has developed steadily over the last 15 to 20 years as R&D in the industry has declined. Up to the mid-1960s, there were many flourishing interactions through technical societies and a myriad of personal contacts. A steady stream of students went from university graduate programs, usually in metallurgy, to stimulating and technically productive careers in the steel industry. However, career opportunities in the industry for materials scientists with graduate degrees are now sharply diminished, and consequently so are the university graduate programs that relate to steelmaking technology and, indeed, to the metals industry in general. This situation will not be turned around easily.

High-Modulus Polymer Fibers

The recent development and commercialization of high-modulus polyethylene (PE) fibers well illustrates the interplay between industrial and academic laboratories, as well as the transfer of knowledge and technology between laboratories in the United States and abroad. High-modulus and high-strength polymer fibers require that the molecules be aligned along the axis of the fiber. It was first noticed in a European university laboratory that, upon stirring a PE solution, fibrous crystals formed on the stirrer. This was in contrast to the usual appearance of folded chain crystals in solution. These fibrous crystals could be processed to form fibers with a high modulus.

It was then shown in a second European university laboratory that PE could be drawn to a high-modulus fiber in the solid state. Special drawing conditions—slow drawing rates at temperatures above those at which molecular segments could move through the crystals—were required to produce high-modulus fibers. In that processing regime, the modulus of the fibers increased with draw ratio. Molecular chain entanglements are an important consideration in that they are responsible for holding the fibers together laterally. The resulting modulus may be as large as one-fourth the modulus of diamond.

The solid-state process was developed further in an American university laboratory, where a hydrostatic extrusion technique was developed to produce high-modulus PE fibers. Very high modulus fibers were produced by appropriate combinations of temperature and extrusion ratio. The extrusion technique was somewhat faster than the drawing method.

The importance of the entanglements was emphasized by the next development, which occurred in a European industrial laboratory. The density of entanglements and the morphology were optimized by converting the polymer solution to a gel before solid-state processing. Gelling sets the molecular topology of the chain entanglements. High-modulus fibers are produced when the gel is drawn to a high ratio.

The gel-and-draw technique has been commercialized by an American corporation, following an extensive in-house development program. A new polymer fiber with distinctive properties has appeared in the marketplace. The starting material is cheap, the process technology is relatively inexpensive, and the modulus and strength are superior. The fundamental information developed early in academic laboratories was essential to the industrial research, which led to successful development and commercialization of the fibers.

OPPORTUNITIES FOR THE FUTURE

The final section of this appendix is devoted to some selected areas in which research in materials processing might have a significant impact in the not-too-distant future.

Computers, Modeling, and Simulation

The rapid development of fast and versatile computers provides opportunities for significant improvements in materials processing technologies. On-line computational control of process parameters can lead to major improvements in product quality and performance as well as to increased efficiency and reduced costs. The union of computers and processing technology already has appeared in some industrial sectors, and greater effort is anticipated in the future. Processing technologies as different as steelmaking, crystal growth, fabrication of integrated circuits, near-net-shape forming, and rapid solidification might benefit greatly from full use of this capability.

To take advantage of this opportunity, major advances will be required in modeling and simulation, characterization of materials behavior during processing, and sensor technology to monitor process inputs and material response. Materials processing typically involves unusual and sometimes extreme conditions of temperature, pressure, strain rate, flow velocity, or other material or system parameters. The phenomena that occur during materials processing are generally complex and highly nonlinear; they must be modeled with considerable precision in order for the numerical results to be meaningful.

Improved process control requires a detailed knowledge of material and system parameters during processing. Thus improved real-time sensing capabilities are required to monitor parameters such as temperature, composition, and flow speed at many locations in the processing system. A sensor development program is under way in the steel industry at present, but a much larger effort and one directed also to other processing technologies is needed.

Advanced numerical schemes for optimizing materials processing may ultimately be able to include not just the process parameters themselves but also criteria relating to performance and life prediction. Fully integrated simulations of this kind will require better understanding of the relations between processing and performance, and also will require improved methods for nondestructive evaluation and testing.

High-Temperature Superconductors

The unexpected discovery of high-temperature superconductors in some ceramic oxides will surely spark a major new effort in materials processing. In fact, this effort is already under way. At this writing, the knowledge base relevant to these new superconducting materials is growing rapidly, and there may well be further surprises in the months ahead. In any event, the processing of these brittle ceramic superconductors to form useful devices will be a high priority for the materials processing community.

A wide variety of shapes and properties will be required if these new superconducting materials are to play their anticipated roles in microelectronics, power transmission, and transportation. Thin films, interconnects, wires, and cables are a few of the configurations that are of interest. Novel processing techniques may be needed, perhaps analogous to those developed earlier for the A15 superconductors, which, because they are intermetallic compounds, are also brittle materials.

It will be necessary to integrate processing studies with structural determinations and property measurements. The high-temperature superconductors under intense study at present are complex oxides of the perovskite family. Their superconducting properties are strongly influenced by the oxygen concentration and the corresponding concentration and spatial distribution of oxygen vacancies. Processing studies of some subtlety will be required in order to establish the relations between processing parameters, oxygen concentration, oxygen defect characteristics, and optimum superconducting properties.

Ceramic Processing

Ceramics, in general, have been the subject of renewed scientific interest because of their potential advantages for use in microelectronic, structural, and sensing applications. The new superconducting ceramics discussed above are just one part of this picture. The problems that will need to be solved in order to realize these advantages are largely in the area of processing.

The unique properties of ceramics stem from the strong ionic (and, in some cases, partially covalent) bonds between the constituent atoms. This kind of bonding leads to unique combinations of properties, specifically: high strength, light weight, retention of strength and shape at high temperatures, and resistance to corrosion and erosion.

The major use-limiting characteristic of ceramics is their brittleness. The strong bonding and complex crystal structures of ceramics cause them to be essentially unable to undergo plastic deformation by motion of dislocations. Consequently, stress concentrations at natural or artificial flaws cannot be relieved by dislocation plasticity, and catastrophic brittle fracture may ensue. This flaw sensitivity of ceramics is responsible for a severe lack of reliability and reproducibility in their mechanical properties.

The synthesis and consolidation of powders is the usual processing technology for ceramics, although the use of techniques such as sputter deposition and chemical vapor deposition for thin-film ceramics is now an active area. Conventional melting methods, widely used for metals, glasses, and polymers, are generally not useful for bulk ceramics because their stability often leads to vaporization before melting, and substantial volume changes accompany melting when it does occur.

Major developments in processing technology will be required in order to improve the toughness of ceramic materials. New methods of producing fine, pure, mono-size powders should lead to improved sintering characteristics and lower sintering temperatures. In ceramics such as zirconia, phase transformations can be exploited to improve toughness through microstructural manipulation. Improved toughness may also be achieved in ceramic-ceramic composites, where strong ceramic fibers are embedded in a brittle ceramic matrix. The strong fibers act as bridges across cracks in the brittle matrix, thereby increasing toughness. There are many processing technologies that can be explored with the goal of improving the toughness of ceramic materials.

Artificially Structured Materials

The recent development of new materials and materials systems structured on an atomic scale should have many important applications. For example, the processes used for producing artificially structured materials make it possible to combine optically active materials with electronic circuitry in ways that should lead to qualitatively new kinds of optoelectronic devices. Artificially structured materials can be produced by a variety of techniques including molecular beam epitaxy (MBE), liquid-phase epitaxy (LPE), chemical vapor deposition (CVD), vacuum evaporation, sputter deposition, ion beam deposition, and solid-phase epitaxy.

Techniques for growing thin films epitaxially, such as MBE, have been used to produce artificially structured materials with levels of purity and structural perfection that seemed impossible only a few years ago. Layered semiconductor systems with layer thicknesses down to atomic dimensions and with atomically smooth interfaces are now grown. The GaAs-GaAlAs system has received the most attention to date, with structures of widely varying electronic properties produced by control of composition. Carrier mobilities in excess of 10^6 cm^2/V-s have been achieved in systems with layer thicknesses of the order of 10^{-6} cm.

An artificially structured material generally can be expected to exhibit novel and useful properties when the length scale of the structure is comparable to the characteristic length scale of the physical phenomenon of interest. Examples of interesting microscopic length scales include the de Broglie wavelengths of electrons, the wavelengths of phonons, the mean free paths of excitations, the range of correlations in disordered structures, characteristic diffusion distances, and the like. These distances can vary from a few atomic spacings to microns. The least explored area, and the one with the greatest potential interest for processing technology, lies between the atomic and the macroscopic sizes.

To date, most interest in the field of artificially structured materials has been focused on semiconductors, but there are many opportunities for new and useful combinations involving metals, insulators, and even polymers. The processing technologies now available are capable of producing both equilibrium and novel nonequilibrium phases, including amorphous structures and extended solid solutions. The range of possibilities in this area is truly remarkable.

Metals Industry

The problems of the metals industry are generally well known. The technical infrastructure of the industry has eroded in recent years, to the extent that the technical knowledge that exists within the industry is not fully used, and the transfer of technology into the industry from outside sources is gravely impeded. As an example, the titanium and nickel industries lack the resources to implement innovative new processes such as electro-refining of titanium and advanced ingot production methods for nickel-based superalloys.

The needs of the metals industry for research in materials processing center on the goal of becoming economically competitive in international markets. In particular, the steel, lead, and zinc industries need new processes that will allow them to take better advantage of their North American ore reserves;

the steel industry is trying to develop a strip casting technology that will circumvent continuous casting operations and hot strip mill complexes; and the lead and zinc industries need extraction processes that will enable them to recover the silver and other minority constituents of their ore deposits.

The most pressing long-term need of the metals industry, however, is the development of processes for recovering metals from waste materials, and for upgrading vast quantities of currently unusable scrap to allow separation and economical recycling. With the exception of steel, there is a general shortage of high-quality raw materials available to the industry from within North America. Technology for efficiently recycling scrap emerges as the most important process area for R&D in the next 15 years.

In the near term, secondary processing operations such as rolling and shaping need attention throughout the metals industry. Process simulation, modeling, sensing, and control improvements are needed to upgrade quality and reduce costs.

For long-term improvements in product quality, truly innovative research is needed to develop predictive models of metallurgical phenomena and to simulate the properties and performance of finished products. This is a potentially important area for collaboration between industry, universities, and national laboratories.

Specific processing areas of interest to the steel industry include the following: the roles of fluid dynamics, heat transfer, and diffusion in determining the mechanism of hot dip coating; optimization of batch anneal heating processes through the development of models that incorporate nitrogen pickup, energy input, and productivity; application of computer-based expert systems to process control; three-dimensional fluid flow analyses of gas distribution in batch annealing, steel flow inside a tundish or mold, and gas flow inside a blast furnace or a basic oxygen furnace; and solidification models for simulating near-net-shape casting.

Polymer Processing

The remarkable growth of the synthetic polymer industry over the last half-century has been due as much to the development of processing technology as to polymer chemistry. Materials processing will continue to be a key to the development of new polymeric systems with unusual properties.

The properties of polymeric materials depend on organization at all levels—atomic, molecular, supermolecular, and macroscopic. Atomic organization is determined in the chemical reactions in which polymers are grown from small molecules. The organization of the polymer molecules themselves, that is, their state of folding or extension, is determined largely by processing. Molecular organization also depends on the chemistry of polymerization, which determines the molecular weights of individual molecules and the ways in which they are branched and cross-linked. At the super-molecular level, polymers interact to form crystalline, liquid crystalline, or amorphous arrays, or mixtures of those forms. Supermolecular organization also emerges from the ways in which different molecules behave when mixed with each other to form, for example, solutions or dispersions. Macrostructure is determined in shaping and finishing operations. Spinning of fibers, extrusion of films, and various molding operations determine macrostructure and often have a profound influence on the lower levels of organization in the bulk material.

The importance of both polymer chemistry and processing on organization and, consequently, on aggregate properties is well illustrated by the new polyethylene fiber. It is prepared by drawing the ultrahigh-molecular-weight polyethylene from dilute (but supersaturated) solutions. The individual molecules are predominantly extended, rather than folded as is the case with polyethylene processed by conventional spinning or molding techniques. Success in making strong fibers by gel spinning has led to efforts to produce strong films by shear spinning, in this case to produce a material with most of the molecules extended in the plane of the fiber.

Important last stages of reaction chemistry often occur during processing. Much effort is now directed toward controlling and using, rather than avoiding, these late chemical changes. In reaction injection molding (RIM), chemicals are mixed immediately before extrusion as films or fibers, or injection into a mold, with the consequence that polymerization occurs concurrently with the formation of the shaped product. This allows the use of polymers that are not easily processed if they are prepolymerized and can also improve process economics. The control procedures needed with a RIM process are very stringent. Good modeling of the reaction kinetics in a moving stream in which temperature, composition, and viscosity are changing rapidly is a tremendous challenge, but, if well done, such models would be enormously useful for production.

Self-organizing polymers, such as those that form liquid crystals in melts or in solutions, are of increasing interest industrially. Objects made by solidifying liquid crystals tend to be strong and stiff. It is expected that they will be important structural and optical materials, but optimization of their properties will require a detailed understanding and control of processing technology.

A new area of polymer processing is based on a classic metallurgical procedure. It has been found that mechanical working (forging) of polymers at elevated temperatures followed by controlled cooling can increase molecular orientation and therefore increase strength markedly. This emerging field promises to be an arena of intense international competition.

A long-term goal is a process for extrusion of thick polymeric objects with high degrees of internal order and uniform properties. If successfully developed, such a process might produce thick sheets, blocks, and shaped objects with strengths now seen only in fibers and films.

Processing of composites involves all of the problems encountered in processing monolithic materials along with a number of new ones. In particular, the organization of reinforcing material in a polymeric matrix may be both critical and difficult to accomplish by economical practices.

Fiber-reinforced composites are a good example. The term *reinforced plastics* often refers to resin-based materials made stiffer by the inclusion of strong, high-modulus fibers, which raise the modulus and strength of the composite by load transfer from the matrix to the fiber. Chopped fibers are widely used because the processing conditions are then similar to those used for monolithic plastics. A continuous filament is much more effective in reinforcement if it can be arranged properly in the matrix.

The simplest process for manufacturing continuous fibers is hand lay-up. The fibers are arranged in a mold, and a thermosetting resin is poured in. In forming the very high quality laminates used in the aerospace industry, the mixture of fiber plus resin is then cured under pressure. An open-mold process is reasonable for production of large parts that do not require optimum properties, e.g., boat hulls or building panels. In either case, the process is labor intensive and not suitable for economic mass production.

The expanded use of composites in applications such as automobile components is contingent upon major improvements in processing technology. Various processes, including injection molding, are suitable for continuous operation, but the reinforcing materials are restricted to short fibers or particles. Continuous processes that use continuous fibers are under development but require much more effort. In the winding of filaments, continuous strands are impregnated with resin and wound onto a mandrel. The pattern of the "fabric" can be designed to produce strength tailored for the desired application. "Pultrusion" is another continuous process and combines pulling and extrusion. The fibers, in the form of strands, mats, or tapes, are pulled through a resin, through a preformer, and then into a die, where the final product is cured.

The electrical properties of composites also depend on the organization of the matrix and the included material. Polymer-based composites are often filled with metals or ceramics whose electrical properties—conductivity, piezoelectric coefficients, dielectric constant, and so on—differ greatly from those of the matrix. The aggregate electrical properties of the composite are complicated functions of the organization of the included or filler material. Important parameters include the interparticle spacing, the connectivity, and the chemical nature of the interfaces. The theory of these structure-property relationships is primitive, and so is the science of controlling the organization of the electroactive particles by controlled processing.

The science that underlies polymer processing often lags behind technological developments. It would be most useful for further progress if good dynamic process models could be developed based upon chemical kinetics, material properties, and mass transport. The systems are complicated, usually inhomogeneous, solids, viscous liquids, or suspensions of solids in viscous liquids. These systems seem difficult to analyze quantitatively, but the potential rewards for developing a systematic understanding of polymer processing are great.

C

Performance

OVERVIEW

Research in the performance of materials cuts broadly across the entire field of materials science and engineering. Performance-related research ranges from understanding the microstructural response of interfaces between complex solids to predicting lifetimes for structural materials subject to stress or corrosion. Progress in these areas depends on tools and perspectives drawn from all of traditional materials science and engineering plus related parts of chemistry, physics, mathematics, and engineering. Scientists and engineers should strive to broaden both their technical perspectives and their professional interactions in approaching problems in this field.

Those responsible for funding in materials science and engineering, especially in the area of performance, which cuts across an unusually wide variety of disciplines, should deal with the field as a whole, recognizing and fostering opportunities for new fundamental advances across a broad front.

Computational modeling of the complex processes involved in synthesis, fabrication, deformation, degradation, and failure of materials is becoming a central tool in research on performance. The increasing power of computers, coupled with new theoretical developments, now makes it possible to develop quantitative methods for solving some of the major problems in this field. In particular, accurate codes may become available to aid in the microstructural design of complex materials such as composites, or for making reliable predictions of the lifetimes of structural materials in service. Research leading to such developments should be strongly encouraged.

Research in processing, manufacturing, and fabrication of materials should be more fully integrated with research on performance and with engineering design of components or structures.

Industry-university collaboration should be strengthened. This is a particularly difficult challenge because much of the industrial base that traditionally has supported research in the performance of materials is in poor health. Industrial areas to which special attention should be paid in this regard include fossil and nuclear power technology, energy resource extraction, surface transportation, and the metals industry.

THE BROAD SCOPE OF RESEARCH IN THE PERFORMANCE OF STRUCTURAL MATERIALS

Research in the performance of materials seeks to predict and improve the way materials behave in service. The materials may be metals, ceramics, polymers, or any of the various alloys or composites that may be formed by combining such constituents. The example of performance research chosen here for detailed discussion is that of structural materials. One wants to know how such materials respond to stresses caused by service loading, mechanical contacts, or temperature variations; how they react to hostile, corrosive environments; and how they undergo internal degradation, e.g., by diffusive processes. The crucial issues are strength, reliability, durability, life prediction, and life extension. Such issues are relevant not just for the materials that are used in large structures or machines but also for those that form the structural elements in electronic, magnetic, or optical devices.

Research in performance involves all the conventional inputs that contribute to understanding relations between the structure of materials and their mechanical (and often chemical and electronic) response properties. It also involves understanding the means of manipulating structures to achieve desired properties and learning how properties, together with operating conditions, determine lifetimes. Here, structure refers not only to atomic structure of constituents and to the atomic bonding between phases, but also to structure at a variety of more macroscopic levels of organization that prove essential to understanding the strength of materials and their failure in service. These macroscopic structures include grains, preexisting cracks or pores, precipitates and inclusions that may contribute to strengthening but may also introduce cracks or voids, fibers embedded in matrices, and surface films or coatings.

Thus the relevant background for performance research includes not only the quantum mechanical concepts essential to understand matter at the atomic scale, and the thermodynamic, chemical, and kinetic concepts needed for understanding structural transformations, but also the more macroscopic concepts of deformation and transport that are relevant to processes that occur on the larger than atomic scales of materials microstructure. (The application of mechanical principles at such scales, to understand strength, deformation, degradation, damage, and failure processes, is commonly called "micromechanics.") Further, since performance research addresses materials in service, its scope includes engineering disciplines such as structural analysis, fracture mechanics, materials forming processes, nondestructive evaluation, tribology, electrochemistry, and corrosion.

We see therefore that the scope of performance research is unusually broad. Few of its currently active researchers and advanced practitioners are involved in the entire spectrum, and fewer yet have received formal education in that breadth. Yet it is evident that many of the most challenging problems in the area call for individuals who can reach well beyond the conventional disciplinary boundaries of contributory fields such as engineering mechanics, metallurgy, polymer chemistry, and solid-state physics. In short, research in performance is particularly an area in which national needs and priorities must be addressed by the materials science and engineering community as a whole.

NEEDS AND INSTITUTIONAL ISSUES

Some general needs and opportunities for research in the performance of materials are presented below.

Technical and Scientific Requirements

• *Improved instrumentation and experimental techniques.* These are needed to characterize microstructures and, especially, to clarify the dynamics of the complex microstructural mechanisms by which materials deform, degrade, and fracture. They are also needed to detect and characterize defects developed in processing, manufacturing, and fabrication, and to trace the evolution of defects in service applications.

• *Better theoretical understanding of performance-related properties and phenomena, and improved capabilities for quantitative analysis and modeling.* This need exists at all levels including the electronic and atomic, micromechanical, and macroscopic engineering domains. The goals are both improved fundamental understanding and the provision of a basis for more enlightened empiricism.

• *Design of materials microstructures for optimized performance.* Advanced microstructural design relies on progress in the two categories just mentioned. It will be carried out best at laboratories where new ideas regarding relations between microstructure and performance may be investigated in collaboration with innovative research in synthesis and processing.

• *Improved methods for prediction and assurance of lifetime in service.*

Techniques for predicting performance of materials should be based on fundamental understanding of degradation and failure processes, on rational test procedures, and on improved nondestructive evaluation technology to detect and characterize cracking or damage in service. For new components or structures, these methodologies should be part of an integrated engineering process involving not only design to meet performance requirements but also the concomitant modeling of the processing, manufacturing, or fabrication steps necessary to bring a material to service.

It may be noted that these needs require, for their fulfillment, work across a range of materials science and engineering activities that often stretches beyond traditional disciplines. The work goes hand in hand with characterization, synthesis, processing, analysis and modeling, and engineering design.

Education

Neither the current status of education in materials science and engineering nor the priorities for funding are fully suitable for meeting needs in the performance area. For example, individuals active in this area have generally had education in core materials science and engineering, in a mechanics-related engineering field, or, to a lesser extent, in physics or chemistry. The core materials programs sometimes lack emphasis on the theoretical background and techniques of quantitative modeling underlying performance research. Those from mechanics-related engineering programs often have minimal exposure to structure-property relations, techniques of manipulating microstructures, and materials phenomena at atomic or molecular scales. Those educated in physics are usually without exposure to core concepts in performance research such as the mechanics of strength and failure processes, and those in chemistry often consider structure-property relations only in terms of small molecules or, at most, highly ordered crystals, rather than in terms of the multiplicity of solid-solid interfaces pervading most useful materials. These lacks are to some extent inevitable, but it is essential that increasing numbers of capable students be encouraged to bridge traditional areas. They will find much productive work at the interfaces.

SCIENTIFIC AND TECHNICAL ISSUES

The basic scientific issues addressed by research in the performance of structural materials are strength, deformability, chemical degradation, and fracture. The goals are to predict, control, and improve the integrity of materials in service. In the following paragraphs, these issues are discussed from a scientific and technical point of view. The discussion is organized according to the length scales—from atomistic to macroscopic—that characterize the underlying phenomena.

Atomic Scales

Atomistic studies of strength, fracture, and chemical reactions are becoming increasingly important in optimizing the performance of materials. For example, modern metallic and ceramic structural materials, as well as composites and artificially structured devices, typically contain large numbers of interfaces separating grains or phases. Fracture at such interfaces, which is frequently influenced by local segregation of solutes and of environmentally introduced impurities such as hydrogen, is often critical in limiting toughness. New fundamental approaches to alloy design, e.g., identification of favorable trace elements that will improve interfacial toughness, should follow from basic understanding of the atomistics of interfacial fracture. One example, so far understood only empirically, is that the polycrystalline-ordered Ni_3Al alloy normally fractures easily along grain boundaries, but may be made ductile by addition of trace boron atoms, which, apparently, segregate at the boundaries in such a way as to strengthen them. An opposite example is the weakening of quenched and tempered martensitic steels by grain-boundary segregation of trace elements such as phosphorus, sulfur, tin, antimony, and, especially, hydrogen.

Analysis and modeling at the atomic scale should be useful in two primary ways. The first is by providing accurate, quantum mechanical calculations of material properties (interfacial energies, solute binding energies, dislocation core configurations and energies, and so on), which are needed in continuum theories of deformation or fracture but which are not easily or accurately measured. Dramatic improvements in our ability to make such calculations have been achieved in recent years, and the impact of these developments is only now beginning to be felt. The second way in which atomic-scale modeling is becoming useful is in exploring the dynamics of complex, many-atom processes. Such calculations necessarily involve departures from rigorous quantum mechanical principles, for example, the use of pair potentials or a modern improvement like the "embedded atom method." Examples of processes for which dynamic modeling is currently being attempted include fracture at a crack tip, chemical bond formation during separation of fracture surfaces, the motion of dislocations, interaction of glide and grain-boundary dislocations, nucleation of martensitic transformations, and entanglements of long-chain molecules during deformation and crazing of polymeric materials.

Micromechanics of Strength and Fracture

Studies of fracture by electron microscopy reveal that the behavior of crack tips is influenced by microstructure at scales well above atomic (e.g., grains, inclusions, and fibers) and often involves, at least in metallic and nonglassy polymeric materials, substantial amounts of local plastic flow. The study of these processes will be aided, often in critical ways, by what is learned atomistically. Nevertheless, such studies depend principally on the largely independent theoretical framework of solid continuum mechanics, specifically, crack theory, dislocation theory, and the constitutive descriptions of plastic, viscoelastic, or viscoplastic flow. Studies in this latter domain will be critical in designing microstructures to optimize the performance of materials.

The domain of "micromechanics" is thus the study of processes that occur in solids on length scales roughly of the order of microns. These scales are much larger than atomic, thus justifying the use of continuum theories, but they are much smaller than the bulk scales ultimately of interest in engineering design. The important problems are to understand processes pertaining to strength and fracture in heterogeneous microstructures.

The conceptual starting point for research in micromechanics is the idea of micron-sized defects that, ultimately, may lead to macroscopic failure of solid objects. These defects are produced in many ways. Sometimes microcracks or cavities are introduced into newly formed materials by synthesis and processing; sometimes they are generated during deformation in secondary processing, manufacturing, fabrication, or service. In metallic systems, nucleation of precipitates typically is accompanied by extensive plastic strain. In ceramics, nucleation of new phases may be activated at incompletely sintered zones or by local stresses generated by elastic modulus or thermal expansion mismatch. Suitable theories to explain the separate effects of macroscopic stress and inelastic strain on nucleation do not exist.

At high temperatures, stress-assisted and transport-assisted cavity nucleation becomes possible. This process occurs along interfaces, often at the sites of inclusions along grain boundaries, and provides the cavities that enlarge to failure in the high-temperature creep rupture of metals and ceramics. Here, too, there are problems in explaining observed behavior in that stresses predicted for nucleation according to the standard model of vacancy condensation generally exceed stress levels inferred experimentally.

Once nucleated, the growth to linkage of microcracks or cavities, causing final fracture, is usually an extremely complex process in multiphase alloys and composite materials. It can also be complex in single-phase solids. The complexity induced by microstructural alterations can frequently be beneficial, and progress toward the understanding of complex fracture processes can pay handsomely in improved properties of materials.

Toughened Ceramics

One example of recent progress in micromechanics is the toughening of ceramics. Ceramics normally are extremely brittle, and yet many are attractive for applications because of their strength at high temperatures (e.g., for heat engines), low thermal conductivity, low density, hardness (for cutting tools), usually low coefficient of friction, and corrosion resistance (for coatings). Routes to toughening of ceramics have been discovered empirically but are now to some extent explained theoretically and are currently being exploited in materials development. These routes include stress-induced phase transformations within initially constrained inclusions (e.g., partially stabilized zirconia), the introduction of ductile metallic inclusions (cermets) that can attract and pin macrocracks, and whisker inclusions that can slide in relation to the matrix and bridge the crack surfaces.

Ductile Rupture

Another example of the use of micromechanics is in studying the mechanisms that lead to rupture in metals used for structural purposes. Generally, such metals are chosen for their ductility. They are thought to fracture by the nucleation of cavities that grow by ductile deformation. Currently promising approaches to understanding the growth to coalescence of these cavities are based on models of continuum plasticity. The key problem is to understand the overall deformation of a porous solid whose material elements obey nonlinear elastic-plastic constitutive relations. Of interest is the growth of such voids to coalescence with their neighbors and, especially, the local instabilities in deformation between pairs or among clusters of voids, which can lead rapidly to macroscopic fracture.

Ductile-Brittle Transition

Some classes of metallic alloys, including the carbon steels used extensively in large structural applications, show transitions from ductile response to low-energy cleavage with decrease of temperature, increase of loading rate, or long-term exposure to neutron irradiation. While such transitions have been known and characterized experimentally for many years, their fundamental explanation is incomplete. The factors that induce embrittlement are also those that increase the resistance to plastic flow, and some degree of understanding has been obtained on this basis. For example, increased strength causes increased stress ahead of a crack or a sharp notch. This stress acts to nucleate running cracks at brittle phases, such as carbides in steel, and also makes it possible for the crack to continue in a brittle cleavage mode across the adjoining metallic grains. However, what is still poorly understood despite impressive recent advances in theoretical modeling is how the strain rate and temperature dependence of plastic flow, together with specific loading or brittle phase crack-nucleation conditions, govern whether a potentially cleavable lattice will in fact sustain such a brittle cracking mode.

Deformational Instabilities and Pattern Formation

At somewhat more macroscopic levels in the analysis of ductile materials, problems arise in understanding the instabilities and patterns that occur in large plastic deformations. An important example is the onset of shear localization. Sometimes this phenomenon is triggered by incipient cavitation at inclusions, but it may also result from an intrinsic instability of the multi-dislocation motion itself. Shear localization frequently leads to profuse voidage within the shear zone and to ductile rupture. Some important theoretical advances have been made in modeling such localization phenomena, e.g., as instabilities in nonlinear elastic-plastic or viscoplastic behavior. However, even within this approach, there does not yet exist a suitable method for analyzing shear localization in strongly nonhomogeneous deformation fields such as those that exist at the tips of macroscopic cracks.

Shear localization strongly limits ductility. In some high-strength alloys it takes place at macroscopic crack tips and leads to low-energy zigzag fracture paths, consisting of one shear localization-induced rupture followed by another. On the microscale, these fractures show extreme ductility; macroscopically, they are extremely brittle in character. Also, while fracture by ductile void growth to coalescence can lead to substantial macroscopic strain at fracture, the actual strain at fracture is often sharply reduced by localization. For example, it is common that well before voids generated from large impurity inclusions grow to coalescence with one another, localized shear bands develop between them and abruptly terminate the process by triggering profuse voidage from families of smaller precipitates within the band.

In addition, there is evidence from studies of hydrogen in localization-prone steels that hydrogen can enhance flow localization. This phenomenon too lacks suitable theoretical explanation, but it is important as a mechanism of hydrogen embrittlement other than by the direct degradation of cohesion along interfaces.

Other types of pattern formation for which present understanding is inadequate include the development of patchy slip textures, where active slip systems vary substantially from region to region of a crystal under nominally identical stresses. In cyclic plastic deformation, patterns of "persistent slip bands" are often observed to take up most of the macroscopic straining. Such bands lead to stress concentrations at material surfaces and are important mechanisms for fatigue crack nucleation.

Contact and Wear

Modern phenomenological descriptions of wear in sliding and rolling contact postulate microstructural mechanisms that are similar to some of those just discussed in connection with strength and fracture. The study of friction and wear is a major part of tribology (which also includes lubrication and machine dynamics). While traditionally studied independently of other work on the micromechanics of strength and fracture, the same basic concepts and experimental techniques have proven essential.

For example, repetitive rolling or sliding contact in ductile systems results in repetitive plastic shear parallel to the contact plane in subsurface material. This leads to localized shear, crack nucleation, and, ultimately, the flaking-off of wear particles. Neither the microstructural alterations caused by wear nor the basis for environmental sensitivity of these processes is well understood at present. Progress in this area should lead not only to better design of wear-resistant structural materials but also to better understanding of wear inhibition by surface coating or ion implantation. In brittle systems, sliding contact produces arrays of microcracks, which ultimately join to form wear particles. Similar processes occur in erosion. They too involve complex mechanisms that are not well understood at present.

Polymers

Many ductile polymeric materials exhibit phenomena analogous to those observed in metallic systems. For example, failure during ongoing deformation is known to be caused by nucleation, growth, and coalescence of voids. Instabilities such as shear localization are also observed. Fractures in ductile systems such as polyethylene often are preceded by the formation of extensive "craze" zones in which the entangled molecules, initially without preferred orientations, form extended zones of fibrils oriented parallel to the direction of maximum stress. Significant improvements of toughness of brittle glassy polymers such as polystyrene and polymethyl methacrylate have been achieved by microstructural manipulation, particularly by blending them with elastomeric second phases, which arrest incipient crazes.

Polymers are being used increasingly in structural applications as matrices for composites, as adhesive joints, and, in some cases, as strong materials (such as aromatic polyamide polymers) that can be used themselves as fiber-reinforcing phases, e.g., in concrete. In order to assess the performance of such composites, it will be important to understand the relatively slow processes that control the degradation of the polymeric components, for example, viscous flow, moisture uptake, and radiation-induced molecular rearrangements.

In summary, improved fundamental understanding of micromechanical properties is going to be needed in order to design at the microstructural level new materials that will meet advanced standards of performance. The class of materials for which rational microstructural design looks promising is remarkably broad. It includes very fine grained single-phase materials, dispersion-strengthened solids, intermetallic compounds, toughened ceramics, polymer alloys and blends, composite materials of all types, thin films, and layered solids. It also includes high-performance concretes with relatively impermeable and hence degradation-resistant pore structures. There is clearly a large amount of work to be done.

Crack Growth, Degradation, Damage, and Life Prediction

The domain of macromechanics includes processes that can be described on macroscopic length scales, that is, scales comparable to the sizes of the devices or structures whose performance is being evaluated. Some topics in research on the performance of materials that lie in this domain are discussed below.

Fatigue

Most service conditions involve time-dependent stresses. Often the most critical questions for evaluation of performance and lifetime relate to the behavior of materials exposed to cyclic loading. Here, just as in the topics discussed in the section "Micromechanics of Strength and Fracture," there is no shortage of complexity. The basic mechanism of fatigue crack growth is an irreversible process in which the crack tip opens under increasing load but does not return fully to its original configuration when the load is released. It turns out that the extent of crack tip opening and the degree to which it can be reversed by closure slip and rewelding are sensitive to the environment. Further, the loading actually sensed at the crack tip differs from that for an idealized crack with traction-free surfaces because the crack walls near the tip come into mechanical contact with each other during the part of the cycle when the load is decreasing. Closure is promoted by protrusions left on the fracture surface by deformation markings from prior cycles and by irregularities of the fracture path. Further, the chemically reacted state of the newly formed fracture surface, i.e., an oxidized or corroded layer, also affects the geometry of the surfaces and the tendency for closure.

Careful experimentation has shown that this picture of crack closure can explain qualitatively the observed dependence of growth rates on load level, crack depth, and external environment in the sense that growth per cycle correlates with the range of near-tip loading over which there is no crack closure. However, a basis is still lacking for fundamental prediction of crack growth rates or for identification of microstructural alterations that might improve fatigue resistance. Also, the methodology for accurate prediction of lifetime in service remains incomplete because of our still limited understanding of crack nucleation and crack growth in the short crack regime where crack depths are comparable to or only a few times larger than microstructural sizes such as grain diameters.

Corrosion and Environmentally Assisted Cracking

Environmental chemistry strongly influences crack tip mechanisms in both fatigue and sustained-load growth (e.g., stress corrosion cracking). The mechanisms are varied, and our understanding of them is highly incomplete. They include the hydration of silicon-oxygen bonds of silica glasses in moist environments; the successive formation and rupture of oxide films at crack tips, usually along grain boundaries, in many metallic systems; and, as in the high-strength steels, the evolution of hydrogen in aqueous surface reactions, which promotes hydrogen embrittlement.

Nearly all service applications of materials in hostile environments involve large fluctuations of loading. As a result, in many applications, it is necessary to predict life on time scales well beyond those that can reasonably be covered by testing. A critical focus for research is on understanding crack growth and other forms of damage in conditions involving interactive fatigue-corrosion or fatigue-creep-corrosion. To a first approximation, pure fatigue crack growth is determined by the number of load cycles but not by their rate of occurrence. Interactive growth rates, on the other hand, are not accurately described as a simple sum of pure fatigue and sustained load rates. It will be essential for progress in developing reliable methods for life prediction that the physical basis of these interactions be clarified. The goal is to provide sound predictions of lifetimes extending, say, 10 to 30 years from short-time tests.

In addition to modeling crack tip separation processes, studies in this area should include the modeling of transport, fluid flow, and electrochemical potential in cracks under steady and oscillating loading.

Oxidation

Research on oxidation should address the critical conditions for transitions between protective film formation and internal oxidation. An important issue is the healing of breaks in a protective scale, which may limit the high-temperature applications of intermetallics such as Ti_3Al. Protective coatings are used for high-temperature oxidation resistance, and research is needed on the mechanisms and kinetics of the breakdown of diffusion barriers between alloys and such coatings.

Transport-related limits to performance also arise with integrated circuits.

These occur as electrotransport, thermal diffusion, and strain-induced transport, and include diffusional phase transformations at ohmic contacts.

Macromechanics of Crack Growth

Other research of critical importance to the prediction of lifetime in service involves the macroscopic mechanics of crack growth and other types of damage, especially under conditions extending beyond those of linear elastic, quasi-static fracture mechanics. Major progress has been made in recent years on the elastic-plastic continuum mechanics of crack tip fields and on the characterization of resistance to quasi-static ductile crack growth. However, ductile crack tip response to complex load histories, as in low-cycle fatigue, as well as crack surface closure effects, remains poorly understood from the standpoint of predictive modeling.

Another important area of continuing research is the unsteady dynamics of crack growth and crack arrest. This area includes the viscoplastic dynamics of cracking in ductile solids, a problem that is critical for understanding ductile-to-brittle transitions. It also includes problems such as the arrest of cracks nucleated in more brittle portions of structures, for example, in the radiation-embrittled region near the inner wall of a reactor pressure vessel. Other problems in the forefront of current research include subcritical crack growth in the high-temperature creep range. There, stress relaxations and redistributions following load transients, as well as the effects on the crack tip of transient creep and stress alterations induced by the crack motion itself, must be understood in order to provide a suitable framework for life prediction. Related issues arise in a form that is still virtually unconfronted in cyclic loading, for example, in creep-fatigue interactions.

Distributed Damage

While macroscopic fracture mechanics has developed according to the concept of a single dominant crack, there are circumstances when a more realistic picture is that of a broadly distributed damage zone. Here, damage refers to a multitude of microcracks or cavities such as occurs, for example, in creep rupture by broadly distributed cavitation. Also, failure of composites under cyclic fatigue loadings typically involves a large number of local cracks, whether in the matrix phase or as isolated fiber breaks, which degrade properties but join together as a throughgoing crack only in the final phase of fracture. A suitable mechanics of damage states needs to be developed, rooted in studies of the micromechanics of failure as discussed earlier, but suitable as a basis for engineering analysis of response to complex stress states and temperature histories in service applications. Situations intermediate to that of broadly distributed damage and of a single dominant crack are frequently encountered. For example, many successfully designed composites and brittle materials such as concrete fracture under rising load with an extensive damage zone at the crack tip. Typically, the macrocrack surfaces remain bridged by incompletely pulled out fibers, or by frictionally restrained aggregate particles, and this provides a significant contribution to material toughness.

At present, empirical relations are used in life prediction studies to describe the rates of growth of cracks or accumulation of damage in terms of load history, temperature, and environment. The complexity of these problems dictates that such empirical procedures will remain prominently in use, but an achievable goal for research is the provision of a more enlightened basis for them. For example, research in macroscopic fracture mechanics in the nonlinear range has identified parameters that, in certain defined circumstances, characterize the severity of deformation near the crack tip and hence serve as the loading variable in terms of which crack growth rate should be characterized in empirical studies. Current approaches to ductile tearing-mode cracking and elevated-temperature creep crack growth provide examples. Also, simple theoretical models of creep deformation and cavity growth, over a broad range of stresses and temperatures, lead to maps of deformation and fracture mechanisms. These maps subdivide the stress and temperature plane into regimes in which one mechanism or another (e.g., diffusive creep versus dislocation creep, diffusive cavity growth versus plastically assisted cavity growth) is dominant. The map concept provides a caution that empirically based relations will, likewise, have limited domains in which they accurately describe deformation, and the maps themselves suggest where to look for those limits.

Nondestructive Evaluation

The aim of fracture mechanics is to predict the growth to failure of cracks or other damage in materials. To be effective for predicting lifetimes, such work should go hand in hand with nondestructive evaluation of materials and structures for defects. Here, research challenges occur in sensor technology for making the necessary measurements, sometimes under hostile conditions and with limited access. Also, research is needed on the quantification of nondestructive evaluation signals so that the information about the state of the material provided by such techniques can be used with confidence in estimating lifetimes.

D

Instrumentation

OVERVIEW

"The whole progress of research shows that discoveries depend on more and more powerful instruments," said Erich Bloch, director of the National Science Foundation, in April 1985 at the tenth annual American Association for the Advancement of Science colloquium in Washington, D.C. A better illustration of his statement could not be found than the development of the scanning tunneling electron microscope, which has enabled materials scientists to "see" atoms on the surfaces of solids. We can now "see" reconstructed surfaces, defects, step edges, and even dangling atomic bonds. G. Binnig and H. Rohrer received the Nobel Prize in 1986 for their development of this new instrument.

Successful development and use of scientific instruments are essential for research in materials science and engineering. Increasingly sophisticated instruments are needed as we push our understanding of materials to more microscopic levels, and as the preparation of new materials becomes more exotic. This field of science uses advanced instrumentation not only to characterize materials but also, in many cases, to prepare new materials. The cost of this equipment is rising rapidly in relation to the overall scientific budget, creating an obvious problem. The development of new instrumentation is not only costly, however; it also requires reliable long-term commitments by both the working scientists and the agencies and institutions upon which they must rely for support. The principal thesis of this appendix is that, within large areas related to materials research in the United States, such commitments have been inadequate to meet present needs and opportunities.

The results of inadequate U.S. commitment to instrumentation in materials research have become increasingly apparent in recent years. The current shortage of state-of-the-art instrumentation for preparing and characterizing materials is a problem that has been discussed by several national study groups and has been addressed by special instrumentation programs in several federal funding agencies. The outward manifestations of this problem are inadequate and obsolete instrumentation in universities and government laboratories, a growing dependence on foreign laboratories for the development of advanced new instrumentation, too small a number of students being trained in the use or development of sophisticated instruments, too few small U.S. companies capable of developing and manufacturing new instruments, and inabilities of larger industrial laboratories to apply modern techniques of measurement science to fabrication and processing.

The committee's survey of the current state of instrumentation in materials science and engineering has led it to the following conclusions:

• The shortage of modern instrumentation in U.S. laboratories is a symptom of a problem deeply embedded in the materials science and engineering community. Too low a priority has been placed on instrumentation, instrumentation development, and measurement science as a whole.

• The characterization of materials and the instrumentation that goes with it are still regarded as routine service functions in many materials laboratories. Scientists working on the development of new instrumentation or measurement techniques are not viewed as being part of the scientific or engineering elite.

• A very large fraction of the worldwide development of new instruments currently takes place outside the United States. As a result, state-of-the-art instruments often are being used in foreign laboratories to prepare or characterize materials long before they appear in the United States. The committee's survey of the instrumentation used, for example, by surface scientists, indicates that there is approximately a 5-year delay between the time a new instrument is announced to the scientific community and the time it is transferred to research laboratories in the United States. It therefore seems possible that a significant part of U.S. materials research will fall some years behind that of our foreign competitors unless we do more to encourage the development of instrumentation in this country.

• The lack of research in instrument development in United States universities has created a shortage of students trained in sophisticated instrumentation.

• Innovative developments in instrumentation do occur in the United States, but our system often does not seem to be capable of sustaining these innovations. A common story in the development of a new instrument is that the first paper to describe a new technique is published by a U.S. scientist, but the real development of the instrument takes place abroad. It then takes approximately 5 years before this equipment is developed by an instrument company (usually overseas) into a commercially available product.

There are many economic and cultural reasons why instrumentation development does not flourish in the United States as well as it does abroad, and most of these reasons are beyond the scope of this report. But there are two problem areas where we have the opportunity to make positive changes: first, in the attitude toward instrumentation in our materials laboratories and, second, in the role played by U.S. national and federal laboratories. State-sponsored laboratories in other countries, especially in Europe and Japan, play a much more important role in instrument development than they do in the United States, and private instrument companies abroad (which are more numerous than those in the United States) work much more closely with the universities and national laboratories. The committee believes that cooperation of this kind can and should be improved in the United States.

If the United States is to stay at the forefront of materials research, then major resources must be devoted to the development of instrumentation, and special attention must be paid to the transfer of innovative instrumentation technology to research laboratories, to commercial instrument manufacturers, and eventually to the industries that will use this technology. In order for this to happen, there will have to be some change in the priorities of the materials science and engineering community.

Unlike other scientific communities such as high-energy physics, astrophysics, and biology, the materials community devotes only a minimal fraction of its resources to the development of new instrumentation. Less than 1 percent of the budget of the Division of Materials Research (DMR) of the National Science Foundation (NSF) is allocated for this purpose; policies in some parts of the Department of Defense (DOD) and the Department of Energy (DOE) reject responsibility for this activity. By and large, novel advances in instrumentation relevant to materials research are not generated by companies or large mission-oriented projects; they are generated by individual scientists with specific purposes in mind. It is this particular kind of inventiveness that must once again be encouraged in the United States.

Programs for the development of innovative instrumentation for materials research should be encouraged and supported at U.S. universities. Funding for such programs should amount to approximately 5 percent of the support for academic materials research as a whole. Proposals from individual investigators should have high priority. Support for multi-investigator research in instrumentation also should be considered.

The national and federal laboratories should begin to play a central role in developing new instrumentation for materials research and should act as the vehicle for rapid transfer of this new technology to the industrial community.

The National Institute of Standards and Technology (formerly the National Bureau of Standards) is expected to play a key role in instrumentation or measurement-science programs aimed at materials science and engineering. Its responsibilities include the development and testing of new instrumentation for basic research and the transfer of instrumentation technology to industry. New resources would be needed in order for this to happen.

There is a real opportunity at present for U.S. policymakers to strengthen both materials research and our system of national laboratories by redirecting the missions of these laboratories. The Packard Committee, which was commissioned by the Office of Science and Technology Policy to review federal laboratories, has asked that the laboratories define their missions more clearly and make these missions different from each other. The present committee recommends that the long-term development of materials-related instrumentation become a more significant part of the new mission of one or more of these laboratories. A very good example of an already existing activity of this kind is the X-Ray Optics Center at Lawrence Berkeley Laboratory. Another example of how state-sponsored laboratories can play central roles in the development of new instrumentation is the German Max Planck institutes.

The interaction between universities, government laboratories, and instrument companies must be enhanced with the objective of creating a healthy and expanding commercial instrumentation community that serves the needs of U.S. materials research. This objective might be accomplished by federal programs in which, for example, manufacturing experts from small instrument companies are encouraged to collaborate with scientists at universities or government laboratories in developing new instrumentation.

PRIORITIES FOR INSTRUMENTATION AND INSTRUMENT DEVELOPMENT IN MATERIALS RESEARCH

In preparing this appendix, the committee asked two different but related questions concerning the priority placed on instrumentation by materials scientists and engineers and the agencies upon which they depend for support. First, what priority is placed on the acquisition of new instrumentation relative to carrying out research with existing equipment? Second, how much emphasis is placed on the development of new instruments as part of research programs?

These two questions seem to be related in the following sense: If a scientific community really believes that instrumentation is important, then it will advocate support for the development of new instrumentation in order to make sure that it has ready access to the most modern equipment. The committee finds that, for reasons apparently having to do with the sociology and reward structure of this field, altogether too low a priority is placed on both the development and the acquisition of new instrumentation.

How do materials scientists spend their money? Statistics from NSF, DOD, and DOE all show that the percentage of money spent for capital equipment by individual investigators at universities ranges from 8 to 14 percent. It may legitimately be argued that this small percentage does not reflect the importance of instrumentation to these scientists. The grants are usually so small that principal investigators really do not have the option of using this money to buy major pieces of equipment. When budgets are cut, priority necessarily must be assigned to people rather than hardware. In recent years, all of the major funding agencies have initiated instrumentation programs to which scientists can apply for support for new equipment. These programs have been adopted as attempts to redistribute the money so that more is spent on instrumentation. One obvious conclusion is that these programs have been made necessary by the fact that materials scientists and engineers have not been willing or able, on their own initiatives, to spend their research funds on instrumentation.

Another measure of attitudes toward instrumentation may be gained by looking at the performance of the NSF-sponsored materials research laboratories (MRLs). These laboratories are block-funded by NSF at levels high enough to permit purchases of major items of equipment, and decisions about how the money is spent are made by the MRL scientists themselves. The MRLs have been operating with some stability for over 25 years, and one of their major features is supposed to be emphasis on the development of central experimental facilities.

During the period from 1979 to 1982, only about 12 percent of the overall MRL budget was spent on experimental equipment. In 1982, the management of this program at NSF recognized that this percentage was too low and took steps to increase it. This was accomplished by taking money out of core support and giving it back as "equipment supplements." The stated goal was to increase the support of instrumentation to 20 percent of the MRL budget by 1984. This goal was finally reached in 1987. Once again, one can see that the materials research community has been unwilling or unable to pay adequate attention to instrumentation and has had to depend on federal funding agencies not only for money but also for help in setting priorities. This problem is clearly related to funding, but the committee believes that it is also, to a significant extent, sociological in that there is a natural tendency in the research community to preserve skilled personnel.

In regard to the question of support for—and interest in—the development of new instrumentation in the United States, the committee has observed that there is relatively little money going into this area of research. The instrumentation program in the Division of Materials Sciences at DOE explicitly excludes support for instrument development. DOD has no documented case of instrument development in materials science being funded through their new University Research Initiatives program. Only the NSF instrumentation programs show money being allocated to development. The instrumentation program in DMR has put $1 million (20 percent of the program's budget) into instrument development—approximately 1 percent of the total DMR budget.

National Science Foundation program officers have argued that, even if the funding agencies were able to supply more money for instrument development, neither the NSF nor the structure (or administration) of the scientific community in the United States would be able to provide the supporting staff necessary to sustain effective programs in this area. Instrumentation development programs do not generally produce short-term results, and they require stable, highly trained supporting staffs. Both of these characteristics are inconsistent with the way in which most academic departments operate in the United States. If it is important (as the committee believes it is) that the materials science community have an active university-based instrument development program, then the attitude of the scientific community toward instrument development must change. This change in attitude must be accompanied by changes in the hiring practices of academic departments, changes in the policies of university administrations regarding support staffs, and the initiation of instrument development programs in all of the funding agencies.

The principal consequence of neglecting instrument development in U.S. universities seems clear: there will continue to be a national shortage of scientists trained in this area. This shortage of skilled personnel will have several further consequences. The time lag between the invention of new instruments and their transfer into U.S. academic or industrial laboratories will increase. The United States will lose out to foreign competition in the commercialization of instrumentation, especially by small, specialized instrument companies. More and more of the instrumentation that is purchased using federal monies will be imported. But by far the worst consequence of neglecting the development of instrumentation in U.S. universities is that, unless some other segment of our scientific community such as federal or industrial laboratories takes new initiatives in this area, our materials research community will fall behind its international competition in its experimental capabilities.

In the committee's discussions with funding agencies and with individual materials scientists, it frequently heard the opinion that instrument development is incompatible with the operation of U.S. universities. The pressures of competing for tenure and promotion make it difficult for young scientists to embark on extended and risky development programs, and the formal educational programs of graduate students are shorter than the lifetimes of such projects. The committee believes that these problems are solvable. Examples of successful instrument development programs exist in state-supported academic materials laboratories in Europe and Japan as well as in a few of our own universities. Moreover, instrument development at U.S. universities has traditionally had very high priority in other fields, especially high-energy and nuclear physics, astrophysics, and biology. The committee sees no fundamental reason why this style of research cannot be accepted more broadly in materials science and engineering.

If instrument development is not taking place in our universities, then the only other possibilities are industrial or government laboratories or instrumentation companies. Historically, the big industrial research laboratories like AT&T, IBM, Exxon, and Xerox have supported instrument development, but this effort seems to be declining. The percentage of our equipment that comes from overseas indicates that U.S. instrument companies are not holding their own with foreign competition, particularly in the area of specialized, one-of-a-kind instrumentation. The question is whether the government laboratories are filling the gap in instrument development for materials research.

EXAMPLES OF INSTRUMENT DEVELOPMENT IN MATERIALS RESEARCH: SURFACE SCIENCE

Materials science and engineering is a combination of many traditional disciplines, ranging from solid-state physics to mechanical engineering; thus the instrumentation needed is quite diverse. In an attempt to understand what is happening in instrument development, the committee chose to focus primarily on one part of materials research, specifically, surface science. Surface science is a particularly crucial and active part of materials research, where, historically, there has been a large effort in instrument development. Therefore a study of the trends in this branch of materials science and engineering might give useful information about the health of the field as a whole. The danger was that by looking only at surface science one might obtain a falsely optimistic signal: surface scientists in the United States are reputed to be overly fascinated by their instrumentation. The committee made its decision to trace the history of the development of instrumentation used to characterize the surfaces of materials fully realizing that certain important instrumentation such as molecular beam epitaxy, which is used for materials preparation, would not be addressed. The findings are summarized as follows:

• Twenty years ago, the United States dominated instrument development in surface science. Prime examples are the Auger spectrometer and the low-energy electron diffraction (LEED) display system. In these examples the universities, industrial laboratories, and instrumentation companies were actively involved in the development.

• In the past 20 years, the United States has almost totally lost the dominance that it once enjoyed in this field.

• Innovative design and unique applications of instruments still occur in the United States, but these accomplishments do not seem very often to result in commercial instruments.

Typically, as is shown below, the first applications of new measurement techniques have been announced by U.S. scientists, but the real development of the instruments has taken place in foreign laboratories. Another common occurrence is that U.S. scientists in fields such as elementary-particle physics have developed experimental techniques that are applicable to materials research, and U.S. surface scientists often have been first to adapt these techniques, but that, in the long run, the U.S. scientists have been overtaken by the ability of foreign laboratories to sustain complicated and expensive instrument development programs.

Brief histories of the development of several surface science instruments are given below. The committee believes that these case histories are illustrative of the general state of instrument development in the United States as summarized in the preceding paragraphs.

Scanning Tunneling Microscope

The scanning tunneling microscope produces an image of atomic structure at a surface by measuring the rate at which electrons tunnel quantum mechanically from the surface to a nearby ultrasharp probe. The instrument was developed at IBM Zurich by G. Binnig and H. Rohrer (Nobel Prize in 1986) and was transferred within 2 years to U.S. laboratories. The rapid transfer of this instrument into many other laboratories was a consequence of the open attitude of Binnig and Rohrer, coupled with the fact that a tunneling microscope is a relatively small and inexpensive instrument. It is important to point out that R. Young's group at the National Bureau of Standards had developed the basic concept of the scanning tunneling microscope in 1972. Their "topografiner" had not achieved atomic resolution by the time that their program was terminated by the National Bureau of Standards, and we shall never know whether it might have done so. The fact that an instrumentation program that was pushing back the frontiers of surface science was phased out at a major, federally operated U.S. laboratory is an especially dramatic illustration of the low priority placed on instrument development in this country.

Double-Alignment Ion Scattering

Double-alignment ion scattering, a technique that was developed at FOM, the state-supported Institute for Atomic and Molecular Physics in The Netherlands, uses the channeling of medium-energy charged particles and the blocking of the backscattered ions to determine atomic structures at surfaces or interfaces. The research group at FOM has published important papers reporting the use of this technique for both basic research on surface melting and more practical investigations of atomic arrangements at silicon-silicide interfaces. It took 8 years after the first paper from FOM was published before a publication based on this technique appeared from a U.S. laboratory. The second and third double-alignment ion scattering instruments in the United States have just come on-line at IBM and AT&T Bell Laboratories. The scientists at FOM have worked closely with High Voltage Engineering Europa B.V., a European company that is now selling these systems.

High-Resolution Electron Loss Spectroscopy

In high-resolution electron loss spectroscopy (ELS), monoenergetic electrons are inelastically scattered from a surface and subsequently energy-analyzed. The characteristic losses seen in the scattered beam measure the vibrational modes of molecules adsorbed on the surface or the phonon modes of the clean surface. The basic concepts of this experiment as well as the instrumentation were developed to study molecules in the gas phase. The first application of this technique to surfaces was presented by F. Propst and W. Piper (University of Illinois) in 1967, but the real development of the instrumentation and the procedure into a useful technique for surface science was accomplished by H. Ibach (West Germany) and S. Andersson (Sweden). One of the most demanding applications of this instrument is the measurement of surface phonon dispersion on clean or adsorbate-covered surfaces. Ibach's group reported the first phonon dispersion curve in 1982, and it was 4 years before a phonon curve was measured in the United States (by L. Kesmodel at Indiana University). The instrument development at the Kernforschungs-sanlage in West Germany by Ibach's group was transferred to Leybold-Heraeus, where a commercial ELS system was built and marketed. Until a few years ago it was the only ELS system that could be purchased. At present, two U.S. companies are marketing ELS instruments.

Angle-Resolved Photoemission Using Synchrotron Radiation

Angle-resolved photoemission has become the primary tool for measuring electronic properties both within the bulk and on the surfaces of solids. U.S. scientists led in the development of this technique. In 1977, an analyzer was built by scientists at the National Bureau of Standards (NBS) and the University of Pennsylvania using electron optics developed at NBS for studying gas-phase systems. In 1979 IBM introduced a new two-dimensional display system, and in 1982 a high-resolution instrument with variable momentum resolution was introduced at Bell Laboratories. Yet, in spite of these U.S. developments, all commercially available angle-resolved analyzers are now made and marketed in Europe.

Auger Spectroscopy

Auger spectroscopy and scanning Auger spectroscopy are the standard tools used to determine chemical compositions of surfaces. The development of Auger spectroscopy by the PHI Corporation is probably the greatest success story of U.S. surface instrumentation. Undoubtedly, the success of PHI owes a great deal to the training that W. Peria gave that company's founders when they were students at the University of Minnesota. Peria ran a surface science laboratory where every student learned about designing, building, and testing new instruments. The development of the Auger spectrometer by Palmberg was begun at Cornell, where he was a postdoctoral student, but was brought to fruition at North American Rockwell. Such developments seem increasingly difficult in the present scientific climate.

Low-Energy Electron Diffraction

The diffraction of electrons from surfaces was observed in 1927 by Davisson and Germer, but it was Germer working at Cornell with support from the Varian Corporation who developed the LEED display system that is used today in nearly every surface chamber. Since Germer's work on the LEED display system in the late 1960s, most of the advances in LEED have originated in Europe. Henzler's group at Hannover, West Germany, has developed a high-coherence, spot profile system capable of resolving surface features on a scale of 1000 angstroms. This instrument is being marketed by Leybold-Heraeus (West Germany). Mueller's group in Erlangen, West Germany, has developed new LEED optics and a very fast data acquisition system that allow them to study diffuse low-energy scattering from surfaces as well as scattering by adsorbates that are damaged by the beam in conventional LEED systems. The most up-to-date LEED system is manufactured and sold by Omicron (West Germany).

Electron Microscopy

Two examples drawn from electron microscopy illustrate many of the points made above.

To be useful in surface studies, an electron microscope must keep the specimen clean to ultra-high-vacuum (UHV) standards, it must include various capabilities for treating the surface within the microscope chamber, and it must also be able to provide some tools for surface characterization. The first electron microscope to address these issues effectively was described in 1978 by Takayanagi and his collaborators. This group was working at the Tokyo Institute of Technology as part of a long-running program of studies of epitaxial film growth led by Goro Honjo. Since 1978, this group has published extensive studies of semiconductor and metal surfaces, attacking problems such as atomic reconstruction of surfaces, domain growth, surface phase transitions, and adsorption kinetics. Over 30 papers in this area were cited in two reviews by Takayanagi and Yagi at the 1986 International Congress on Electron Microscopy in Kyoto, Japan. In the United States, the first similar publications date from 1983 (P. Petroff and Wilson), that is, 5 years after the first Japanese work. The first commercial electron microscope with UHV capability was announced by JEOL (Japan) in 1986 and is now available for purchase by U.S. laboratories. Assuming typical instrumental start-up problems, the committee expects that increased U.S. research in this area will begin to appear in the next year or so, about 10 years after the first paper was published by the Tokyo group.

The second example, scanning transmission electron microscopy (STEM), was developed in the United States by A. Crewe and his co-workers. The crucial breakthrough was the invention of an electron gun that had sufficient brightness to achieve high resolution in a scanning mode. Effective use of high-performance electron lenses based on the work of West Germans and, of necessity, UHV techniques resulted in development of a scanning electron beam system capable of focusing a current on the order of 1 nA into a spot less than 5 angstroms in diameter. Many new imaging techniques were made possible by use of inelastically scattered electrons, emitted x-rays, and various forms of the elastically scattered beam. Development of this instrument was initiated by Crewe, an accelerator physicist and then the director at Argonne National Laboratory, and was continued by Crewe and his students after he moved to the University of Chicago. Funding of the development of the instrument was initiated through the director's discretionary fund at Argonne and continued through the National Institutes of Health and the Biological and Radiation Physics section of DOE. Again, this instrument took about 10 years to develop. It was picked up for manufacture by VG Instruments (a British company), which has sold over 20 instruments in the United States, all of which are being used for materials science. Ironically, in view of the initial funding sources, the committee knows of only one home-built instrument in the United States, at Brookhaven National Laboratory, which is being used for biological research. Current developments in this area are being funded by the DMR at NSF. This example illustrates a commitment that, in this case, has had great benefit for materials research. It also illustrates the ability of U.S. scientists to be innovative in this area if the resources are available. Finally, the project was undertaken by an accelerator physicist; it seems unlikely that it could have been started by a materials scientist under today's circumstances.

Low-Energy Electron Microscope

The low-energy electron microscope, which has just recently been demonstrated by Bauer and his co-workers in Clausthal, West Germany, uses a field emission source to create a very bright beam that is collimated at high energy and then decelerated before being back-diffracted from the sample. The diffracted beam is magnified and then used to form an image of the sample. Such images show great promise of revealing defect structures of surfaces.

The development of this microscope took Bauer nearly 20 years. The original program began in the early 1960s while Bauer was working at the Michelson Laboratory in China Lake, California. After approximately 4 years of work, the program began to be phased out by the management because of the lack of published results. Bauer then moved himself and the uncompleted microscope to Clausthal in 1969. Progress in Clausthal was very slow for some years. Finally, in 1975, Bauer and Telieps succeeded in making this complicated microscope function. During the late 1970s, Bauer received very little support from the West German government because a referee had reported that the instrument would never work and that the project should never have been started. It was possible for Bauer to complete the microscope only because, as a chaired professor in a West German university, he had considerable support of his own.

In Bauer's own words, "The success of such a project depends one hundred percent on having the right people and the necessary support at the right time. If one of these conditions is not fulfilled, the project can last forever or fail completely. I was lucky to have fallen into the 'last forever' category."

Spin-Polarized Measurements

Spin-polarized photoemission measures the energy distribution of the spin-dependent states, which determine the magnetic behavior of materials. This class of measurement techniques was first demonstrated in Switzerland in 1969. It was extended by using synchrotron sources and photoelectron energy analysis about 10 years later at a West German government laboratory with an excellent support staff and large financial resources. This instrumentation-intensive experiment is only now being undertaken in the United States by an NSF-supported materials research group consisting of nine institutions including government, university, and industrial laboratories. If this group is successful, it will publish its first paper nearly 10 years after the first West German publication.

In a related development, a qualitative improvement in techniques for producing spin-polarized electrons occurred in 1974 with the invention at the Eidgenossische Technische Hochschule in Zurich of a source using photoemission from GaAs. This source was subsequently developed there and in West Germany, and in the United States at NBS and at the Stanford Linear Accelerator (SLAC). At SLAC, it was used in the landmark measurement of parity nonconservation in high-energy inelastic electron scattering. Despite the very successful application of the GaAs-polarized electron gun in materials studies such as spin-polarized electron scattering and spin-polarized inverse photoemission studies of surface magnetism, few groups are using this device because of its lack of commercial availability. This situation is expected to change, as PHI Corporation is now licensed to manufacture such polarized electron sources.

Spin polarization analyzers have traditionally been large, cumbersome devices operating at energies of 100 keV. More compact analyzers recently have been developed at Rice University, and small low-energy analyzers have been developed in West Germany and at NBS. The NBS spin analyzer has been used on a scanning electron microscope to measure the spin polarization of secondary electrons emitted from a magnetic material and thereby obtain images of magnetic microstructures. A prototype commercial instrument of this type has also been built and tested by PHI and is expected to reach the market in the near future.

Field Ion Microscope and Atom Probe

The field ion microscope and the atom probe, which were developed by Mueller and his collaborators at Pennsylvania State University, produce images of individual atoms at the tips of very sharp needles. They provide good examples of how instrument development can thrive at a U.S. university and how the students in such a program can transfer the technology to other laboratories.

Beam Scattering

The U.S. scientific community has a distinct advantage over its foreign competition in beam scattering, especially in the area of chemical physics. The use of molecular beams to study the dynamics of chemical reactions at surfaces is yielding valuable information about sticking coefficients and state-selective adsorption and desorption. More generally, it is helping physical chemists discover the pathways by which reactions take place. In contrast to the situation in chemistry, however, physicists in the United States are far behind research groups in West Germany in the use of these techniques. Exciting new surface physics is coming from elastic and inelastic scattering experiments, for example, the soliton-like reconstruction of the (111) surface in gold, and the observation of a soft phonon on the (100) surface in tungsten that might be related to a reconstruction. At present, U.S. experimental physicists are just watching.

E

Analysis and Modeling

OVERVIEW

This appendix focuses on some recent theoretical and computational developments that are changing the very nature of materials science and engineering. Two complementary forces are driving these changes. First, there is the unprecedented speed, capacity, and accessibility of computers. Problems in mathematics, data analysis, and communication that seemed untouchable just a few years ago now can be solved quickly and reliably. Second, there is the growing complexity of materials research. The latter change has occurred in large part because we now have instruments with which to make highly detailed and quantitative measurements and we have the computational ability to deal with the resulting wealth of data. Complementing these technology push factors is the pulling force of the technological demand for increasingly complex materials.

Underlying all of these developments are advances in our theoretical understanding of the properties of materials and in our mathematical ability to devise accurate numerical simulations. In short, materials research is evolving into a truly quantitative science.

Analysis and modeling in materials research traditionally has been divided into roughly three different areas of activity—areas that can be characterized by the length scales at which the properties of materials are being considered. The most fundamental models, those used primarily by condensed-matter physicists and quantum chemists, deal with microscopic length scales, where the atomic structure of materials plays an explicit role. At a more phenomenological level, much of the most sophisticated analysis is carried out at intermediate length scales, where continuum models are appropriate. Examples of topics in the latter category include fracture mechanics and microstructural pattern formation in alloys. Finally, there is work at macroscopic length scales, in which the bulk properties of materials are used as inputs to models of manufacturing processes and performance. Historically, research in each of these three areas has been carried out by separate communities of scientists—applied mathematicians, physicists, chemists, metallurgists, ceramists, mechanical engineers, manufacturing engineers, and so on. One of the committee's principal theses is that the distinction between these areas of research is properly being blurred by modern developments.

The principal recommendations of the committee are the following:

• Analysis, modeling, and computation should play a significant role in both the educational and the research components of academic programs in materials science and engineering. Renewed attention should be paid to mathematical analysis (as distinct from—and in addition to—computer programming) in educational curricula.

• New support is needed both to make high-level computational facilities available for materials research and to develop validated data bases, algorithms, and numerical simulations.

• Special attention should be paid to the need for accurate models of nonequilibrium phenomena, particularly processes relevant to manufacturability and performance of materials. Work toward the development of integrated approaches, combining science-based simulations with optimization of features regarding quality and cost, should be strongly encouraged.

PROPERTIES OF MATERIALS AT MICROSCOPIC LENGTH SCALES

Particularly interesting developments of the past few years are apparently feasible schemes for carrying out "first-principles" computations of complex atomic arrangements in materials starting with nothing more than the identities of the atoms and the rules of quantum mechanics. To put these developments in perspective, it will be best to mention some more conventional—and still very productive—approaches to atomistic modeling of materials before turning to this remarkably ambitious new point of view.

Statistical Mechanics

One conventional picture of how huge numbers of atoms collectively determine the bulk properties of materials is the classical statistical mechanics of Boltzmann, Gibbs, Einstein, and others, dating back to the turn of the century. Quantum theory plays only a subsidiary role in this picture. In principle, it is needed to tell us how the atoms interact with one another, but, in practice, these interactions often are replaced by phenomenologically determined pairwise forces between the atoms. Perhaps the most visible success of this classical approach has been the understanding of phase transitions, for example, the crystalline ordering of solids, the compositional ordering of alloys, or the magnetization of ferromagnets. For the most part, however, these are only "in principle" successes, because the details for most cases have yet to be worked out. A promising start has been made in the computation of phase diagrams for multicomponent metallic alloys. (Here the quantum theory of metallic binding does turn out to be of practical importance, but the basic statistical methods remain relevant.)

Research in the area of phase transitions was placed on a much firmer footing in the mid-1970s, when the classic problem of critical phenomena was solved by means of what has come to be known as the "renormalization group" method. With the new understanding that has emerged from this theoretical development, we are now able to classify a wide variety of different kinds of phase transitions and to determine what analytic or numerical approaches might be appropriate for carrying out accurate calculations in various cases of practical interest.

Systems Far from Equilibrium

The remarks in the preceding paragraphs pertain only to what are known as "equilibrium states of matter," that is, to states of matter that have been allowed to relax all the way to thermal and mechanical equilibrium with their surroundings. Many of the most fundamentally challenging and technologically important problems, however, have to do with systems that are far from equilibrium.

For example, the distribution of the chemical constituents in metallic alloys or multicomponent ceramic or polymeric materials is usually very far from what it would be if those materials were allowed to relax by annealing or indefinitely long aging. Thus the problems of predicting relaxation rates and the structural changes that materials undergo on their way toward equilibrium are of great practical importance if one is interested in either the manufacturability or the performance of such materials. Another example occurs during the processing of materials. In processing, substances are almost always driven away from their states of equilibrium. In casting, welding, extrusion, melt spinning, and so on, complex patterns are being made to emerge from relatively simple structures, which means that the natural trend toward equilibrium is actually being reversed. Clearly, the standard techniques of equilibrium statistical mechanics are not adequate in such cases.

Much progress has been made during the past decade or so in the theory of nonequilibrium processes and in techniques for modeling such processes numerically. But many deep problems remain unsolved. A prime illustration of the unsettled state of this field is that we do not yet have a satisfactory way of characterizing the intrinsically nonequilibrium, amorphous—i.e., glassy—states of matter.

Quantum Mechanical Calculations

The modern quantum theory of the structure of materials has its origins in the calculation of the cohesive energy of metals by E. Wigner and F. Seitz in the 1930s. With the advent of large computers during the past two decades, such calculations have achieved quantitative predictive capabilities when applied to regular (or very nearly regular) crystalline solids. Recent developments open the possibility of similar accuracy in describing irregular configurations such as crystalline deformations near defects, surfaces, or grain boundaries. It is even possible that the new methods will allow studies of metastable or strongly disordered states of matter.

In order to predict the structure of a solid, in principle, it is first necessary to calculate the total energy of the underlying many-body system of interacting electrons and nuclei for an arbitrary configuration of these constituents, and then to find the specific configuration that minimizes this energy. A typical computation of the kind that has been tested carefully during the past decade might proceed by, first, fixing the positions of the ion cores, then using what are known as self-consistent "density-functional" and "pseudopotential" methods to find the electronic ground-state energy in this configuration, including ion-ion interactions to compute the total energy, and finally comparing this energy with that of other configurations in order to determine the equilibrium state of the system as a whole. In recent applications of this technique, total energy differences between alternative crystalline structures have been obtained accurately to within a few tenths of an electron volt per atom, structural parameters to within tenths of angstroms, and bulk moduli and phonon frequencies to within a few percent. Note, however, that the method described above pertains only to zero-temperature ground states of regular crystalline arrays of atoms and not to irregular configurations or to alternative phases that might occur at higher temperatures.

Very recently, new methods for performing ab initio total energy calculations have been suggested that provide a novel way of carrying out the above procedure and that also can deal with arbitrary configurations of fairly large arrays of atoms—50 to 100 atoms in a supercell geometry using currently available computers. The basic idea is to minimize the total energy of the system by allowing both the electronic and the ionic degrees of freedom to relax toward equilibrium simultaneously. The useful computational scheme is known as "simulated annealing," a recent development in mathematical optimization theory that has been borrowed from statistical mechanics. (Note the profitable two-way interaction between materials science and other areas of research.) In effect, the system is made to find its minimum energy accurately and efficiently by cooling from a state of high temperature.

If the new ab initio methods can, in fact, be made to work with the anticipated precision, a wide range of materials problems will become open for quantitative investigation. The following examples should provide some sense of the scope of these opportunities.

In the area of surface and interface science, there is interest in predicting the electronic and geometric structure of clean surfaces, grain boundaries, and adsorbed and chemisorbed surfaces. Questions include: How and why does a surface or grain boundary reconstruct? What is the nature of the atomic relaxations? Where might chemisorbed atoms and molecules be attached on a reconstructed surface? What are their binding energies? What are the effects of steps, defects, and impurities? Quantitative answers to such questions now appear to be obtainable, and some relevant investigations have been started. For example, it recently has turned out to be possible to perform an ab initio calculation of the atomic positions at a twist grain boundary in germanium. In comparison to what would have happened if the atoms had been kept frozen in their bulk crystalline positions, the relaxed configuration determined computationally exhibits substantial distortions and the formation of many new covalent bonds.

A closely related area involves interfaces between chemically different substances and the manufacture of artificially structured materials. Here the interest lies in constructing heterojunctions and superlattices of semiconductors and metals in order to obtain special electronic properties. Overlayers and material sandwiches can lead to new structural phases, new electronic states, and new magnetic states. Quantitative predictions concerning interfacial bonding and electronic and magnetic structure of ideal interfaces would be extremely valuable. Present modeling limitations seem primarily to be associated with the long-range stresses and the misfit dislocations that are generated by mismatch of lattice constants across interfaces. Here one can see the need for combining atomistic calculations—the interface structure—with continuum theories—elasticity, plastic deformation, and so on—in a way that may well characterize much of the future work in this area.

Finally, the committee suggests that there is a major opportunity for the application of ab initio methods in the modeling of nonequilibrium processes. For example, in epitaxial growth using molecular beam or chemical vapor deposition techniques, it might be very useful to model the kinetics of atoms interacting with a substrate during growth. One could imagine that the new "annealing" algorithms would be suitable for modeling a wide variety of such growth processes at the surfaces of materials.

CONTINUUM MODELS OF THE PROPERTIES OF MATERIALS

In this section attention is focused on problems in analysis and modeling in which the relevant length scales are on the order of microns or more, that is, much larger than the distances between neighboring atoms. When modeling the behavior of materials at such large length scales, one generally does not need to keep track of the positions of the individual atoms. Rather, it suffices to deal with local average properties—e.g., density, temperature, strain, and magnetization—and to describe the behavior of these quantities by continuum equations in which it is assumed that all variations are extremely slow when viewed on atomic length scales. Thus one uses diffusion equations to describe the transport of heat, composition, or chemical reagents; hydrodynamic equations to describe the motion of fluids; and elasticity theory to relate strains in solids to applied stresses. Of course, many of the ingredients of such models—the transport coefficients, for example—ultimately are determined by fundamental, atomistic principles. But the basic point of view is macroscopic in the sense that it pertains to length scales that are much larger than atomic, and classical in the sense that it makes no explicit use of quantum mechanics.

The advent of the computer has brought about an important change in the perspective from which scientists view continuum analysis. Because continuum models in principle derive from atomistic theories, they often have been viewed as less fundamental, less of a venture into uncharted territory, less apt to produce surprises. It now seems that just the opposite may be true, at least as regards many of the questions that are most relevant to materials research. Now that we actually can explore the consequences of the continuum models—whose ingredients have been known and trusted for decades or longer—we are discovering a wealth of unexpected phenomena and challenging mathematical problems.

In order to describe the implications of some of these recent developments, two broad classes of continuum problems that are part of the traditional core of materials research—microstructural solidification patterns in alloys and fracture mechanics—are discussed below. These are by no means the only areas of materials research where analysis and modeling at continuum length scales are appropriate. Note, for example, the wide variety of materials processing problems in which hydrodynamics is important, or the yet more complicated problems in which fluid motion is coupled to diffusion and chemical reactions. Some of these more complex modeling problems will be referred to toward the end of this section in the discussion of integrated approaches to materials technology.

Microstructures in Alloys

When a molten alloy is solidified by quenching, its chemical constituents tend to segregate. This happens even in situations where equilibrium thermodynamics predicts stable, homogeneous solid solutions. Segregation occurs because, during the solidification process, the liquid and solid phases fall out of equilibrium with each other, and the chemical constituents are rearranged by being driven across the moving solidification front. Thus the last bits of liquid to solidify may be compositionally very different from those that solidified first. The result is often an intricate pattern of cellular or dendritic (treelike) structures, on the scale of tens or hundreds of microns, that are easily observed through an optical microscope.

Control of these microstructural patterns has long been understood to be essential in materials technology. The processes by which the microstructures form are important in determining the grain structure of the solidified material. The microstructures themselves, within each grain, affect the mechanical properties of the material; for example, they may pin or impede the motion of dislocations. They also determine the way the material will behave under further processing such as heating, deformation, aging, or exposure to corrosive environments. The strength of a weld depends on the microstructure in the resolidified material and on microstructural changes in the heat-affected zone. The suitability of a semiconductor crystal for use in electronic devices depends on careful suppression of microstructural solute segregation. Other examples of the technological importance of microstructural properties appear frequently, either explicitly or implicitly, throughout this report.

To the aspiring modeler, the basic ingredients of the microstructural problem may seem pedestrian. Generally, one is being asked simply to solve well-understood diffusion equations subject to apparently simple boundary conditions. The trouble is that the boundaries are moving; in fact, the essence of the problem is to compute their motion. To make matters yet more interesting, microstructural patterns are caused by morphological instabilities of these boundaries. Initially smooth shapes naturally develop grooves and fingers, the fingers split or develop side branches, the side branches split or develop tertiary side branches, and so on. What has been discovered only very recently is that the patterns generated by this process are controlled by an extremely delicate interplay between a basic diffusional instability and a number of ostensibly much weaker effects, most notably surface tension, but also crystalline anisotropy, interfacial attachment kinetics, and even ambient noise. In more technical terms, the instabilities produce intrinsically "nonlinear" behavior, and the weak, controlling effects are "singular perturbations."

The recent developments in this area have been brought about by an interactive combination of mathematical analysis, numerical computation, and careful experimentation. At the time that this report is being written, there is a sense among workers in the field that they may be narrowing in on a new understanding of fundamental principles—for example, that it may finally be possible to compute the growth rates and geometries of dendritic or cellular solidification patterns. But the situation is not yet clear; in fact, the recent history of the field makes it seem likely that further surprises are in store.

Let us assume, however, that the fundamental principles of solidification theory are about to come under control. What happens next? Historically, it has not been easy for advanced concepts in solidification theory to have much impact on applied materials technology, presumably because neither the principles nor the computers have been powerful enough for such work to have quantitative predictive capability. The phenomenologist has managed generally to stay a step or two ahead of the theorist in this field. If this situation is indeed changing, the next step will be to make the new methods of analysis and modeling accessible to process engineers, welding specialists, and the like. Note how difficult this step is going to be. Our aspiring modeler may start out as a computational physicist or a theoretical metallurgist, but he or she will have to master some very subtle mathematical concepts in order to write sensible computer codes, and then will have to learn the language of technology in order to translate numerical results into useful rules of procedure for the processing of materials.

Fracture Mechanics

Research in fracture mechanics is described in Appendix C. The reader should refer to that discussion for a broad view of the importance of this field and a summary of outstanding problems. Fracture mechanics is revisited briefly here in order to focus on certain aspects that pertain specifically to analysis and modeling.

In comparison with the solidification problems discussed above, analysis and modeling in fracture mechanics appear to be considerably more complicated in their physical ingredients and in the variety of their applications. Even the linear theory of elasticity for crystalline solids is intrinsically more complicated for computational purposes than theories of heat transport or solute diffusion. To describe real cracks in real solids, one needs a theory of nonlinear elasticity supplemented by models of plastic deformation and viscoelastic dissipative mechanisms. One also needs to take into account defects, inclusions, grain boundaries, and the overall shape of the object in which the crack is located. It even seems likely that the solution of important problems in fracture mechanics, such as the dynamics of decohesion at a crack tip, will require a quantum theoretic description of atomic bonding of the kind mentioned earlier in this appendix.

Despite its relative complexity, computational modeling in fracture mechanics is currently much more advanced in its impact on materials technology than the modeling of microstructural solidification patterns. This has happened because, as it turns out, much progress can be made in fracture mechanics without directly confronting mathematical problems as difficult as the free boundary problems encountered in solidification. One of the earliest applications of digital computers was the numerical calculation of elastic stresses in irregularly shaped objects subject to forces applied at their (fixed) boundaries. Such calculations, combined with estimates of elastic limits and yield stresses, continue to be immensely important in the design of structural components such as pressure vessels and airplane wings. The same techniques, brought down to micromechanical length scales, are used in fracture mechanics to compute the stress intensity at the tip of a crack and, from this, to determine whether the crack will grow. Such calculations and their extensions are now providing useful, quantitative predictions of the performance of materials. But much important work remains to be done.

A look at new opportunities and outstanding problems in fracture mechanics indicates that the next stage in the development of this field may be just as hard as—and perhaps very much harder than—the problems now (apparently) being solved in solidification. Consider a few examples of important unsolved problems: Microcracks form spontaneously in stressed or heterogeneous materials; ordinarily, it is these defects whose coalescence or growth leads to crack propagation and macroscopic failure. We do not yet have a quantitative understanding of the mechanisms by which microcracks form. Much less is known about the dynamics of crack propagation than the statics of crack initiation. We have no quantitative understanding of what controls the extent of plastic deformation near the tip of a moving crack, how fast the crack will move under given external stresses, or what governs the path along which the crack will move in a heterogeneous medium. Note that all of these problems, like solidification theory, inescapably involve nonequilibrium phenomena; irreversible, dissipative processes play essential roles. Moreover, these are all free boundary problems. Their solutions almost certainly will require detailed analyses of the actual shapes of voids, crack tips, zones of plastic deformation, and so on.

The conclusions to be drawn from these remarks about fracture mechanics are essentially the same as those of the preceding paragraphs regarding microstructural solidification patterns. These conclusions are equally relevant to a much broader class of opportunities for analysis and modeling in materials research, including work at both microscopic and macroscopic length scales. In short, there is a real chance that recent theoretical and computational developments will lead to a new level of predictive capability. However, achieving this level is not going to be easy. The mathematical and scientific problems are hard, and the problems of translating new quantitative capabilities into technologically useful information seem particularly challenging. The latter challenge is a starting point for the remarks that follow.

INTEGRATED APPROACHES TO DESIGN AND MANUFACTURING

To complete this summary of opportunities for analysis and modeling in materials research, ways in which modern computational capabilities might have a direct impact on manufacturing technology are considered. The combination of science-based numerical simulations with new methods for storing, retrieving, and analyzing information ought to make it possible to optimize not just the properties of specific substances but entire processes for turning raw materials into useful objects.

Materials considerations are important throughout the life cycles of most products from design, through manufacturing, to support and maintenance, and finally to disposal or recycling. Significant improvements in quality, reliability, and economy might be realized at all stages of this cycle if quantitative models of processing and performance could be used at the beginning.

Consider, as a simplified example, the design of a turbine disk for an aircraft engine. Under ordinary conditions, the designer starts by attempting to achieve given performance specifications in a way that satisfies a few basic constraints, perhaps minimization of weight in the present circumstance. Fabrication engineers are then asked to see whether the part can be produced. If problems exist, or if the cost appears excessive, some design modifications may be required. Finally, maintenance and inspection specialists are consulted, but it is usually too late by this stage in the procedure for their needs to have a major impact on design. This serial approach emphasizes mechanical aspects of the design at the expense of production and maintenance considerations.

A far better approach is to incorporate all or most of these considerations into computer simulations carried out during the initial stages of the design process. To start, one might examine possible processing paths in order to optimize metallurgical microstructures for different properties in different regions of the component. In the example of the turbine disk, the microstructure could be optimized for creep strength at the rim and for low cycle fatigue and ultimate tensile strength in the bore. The same detailed simulations might also address issues of technical feasibility and economics. Models that relate microstructural properties to processing paths might also be used to examine manufacturing options and even to optimize for ease of maintenance.

It is quite likely that such an integrated approach to materials design eventually can lead to small but significant changes in technology that, in turn, will produce large improvements in performance and cost over the lifetimes of products. It is also possible, because complex problems in systems analysis are involved, that the results of these integrated simulations occasionally will turn out to be very different from what was expected. When that happens, the technological impact is apt to be very great indeed. The limiting factors in this program are the availability of analytic and numerical models and the availability of the specially trained scientists and engineers who are needed to develop these models and bring them to bear on technology.

Index

Note: References pertain to the United States except where otherwise indicated.